PENGUIN BOOKS

GAME, SET AND MATH

Ian Stewart was born in Folkestone in 1945. He graduated from Cambridge University in 1966 with a BA degree in mathematics and obtained a PhD from the University of Warwick in 1969, where he is now a Reader in Mathematics. He has held visiting positions in West Germany, New Zealand, Connecticut and Texas. He is an active research mathematician in nonlinear dynamics and bifurcation theory. He has written popular articles about mathematics for *Scientific American*, *New Scientist*, *Nature*, *Pour La Science*, the *Economist*, *The Times* and the *Guardian* and occasionally contributes to BBC Radio. His books, including translations into twelve languages, number just over a hundred. They include *Concepts of Modern Mathematics* (Pelican 1975), *The Problems of Mathematics*, *The Art of C Programming*, *Does God Play Dice?: The Mathematics of Chaos* (Penguin 1990) and three cartoon books published in French: *Oh! Catastrophe!*, *Les Fractals* and *Ah, les beaux groupes!* He is the European Editor of the *Mathematical Intelligencer* and is on the editorial board of several scientific journals and academic book series, including the A-Level magazine *Mathematics Review*. He is a member of Science Fiction Writers of America, and his SF stories have appeared in *Analog* and *Omni*. He lives in Coventry with his wife, two sons, two cats and a variable number of goldfish.

GAME, SET AND MATH

ENIGMAS AND CONUNDRUMS

———————

IAN STEWART

PENGUIN BOOKS

PENGUIN BOOKS

Published by the Penguin Group
Penguin Books Ltd, 27 Wrights Lane, London W8 5TZ, England
Viking Penguin, a division of Penguin Books USA Inc.
375 Hudson Street, New York, New York 10014, USA
Penguin Books Australia Ltd, Ringwood, Victoria, Australia
Penguin Books Canada Ltd, 2801 John Street, Markham, Ontario, Canada L3R 1B4
Penguin Books (NZ) Ltd, 182–190 Wairau Road, Auckland 10, New Zealand

Penguin Books Ltd, Registered Offices: Harmondsworth, Middlesex, England

First published by Basil Blackwell Ltd 1989
Published in Penguin Books 1991
1 3 5 7 9 10 8 6 4 2

Printed in England by Clays Ltd, St Ives plc

Contents

Preface

A few years ago, Philippe Boulanger asked me to suggest someone to write a "Mathematical Visions" column in *Pour la Science*. That's the French translation of *Scientific American*; Philippe is the editor. I first came across that magazine in my teens, and for me the high point was Martin Gardner's "Mathematical Games" column. When Gardner ceased writing it, the column eventually metamorphosed into A. K. Dewdney's admirable "Computer Recreations". The change is perhaps symbolic of our times. But in France, the idea that computers are here to replace mathematics was resisted, and "Mathematical Games" lived on, in tandem with "Computer Recreations", under the name "Visions mathématiques". That fits my world-view: computing and mathematics have a symbiotic relationship, each needing the other. Anyway, the anchorman of the column had departed for pastures new, and Philippe was looking for a replacement.

Did I know of anyone suitable? Of course I did, and modestly offered my advice. "Me."

He took it – with, I suspect, a few qualms. Two years later, the column has found its own identity and settled into its own style. I write it in English, and Philippe translates it (with considerable skill and also considerable licence) into French. I try to come up with puns that will work in French: for instance, "the twelve games of Christmas". That translates as "les douze *jeux* de Noël", whereas "the twelve days of Christmas" is "les douze *jours* de Noël". And nowadays, whenever I encounter an interesting piece of mathematics, one part of my mind is thinking "I wonder if I could explain that in *Pour la Science* . . . ?" It offers a very different perspective; and on at least one occasion an idea that I had when thinking about "Visions Mathématiques" turned out to be useful in serious research.

Anyway, here it is: *Game, Set, and Math*, a selection of twelve articles that present serious mathematics in less than serious fashion. I've edited them, updated them, and put the English puns back. People sometimes try to sell the idea that "mathematics can be fun". I think that gets the emphasis wrong. To me, mathematics *is* fun, and this book is a natural consequence of the way I approach the subject.

Mind you, I can understand why most people find that statement baffling. To see *why* mathematics is fun, you have to find the right perspective. You have to stop being overawed by symbols and jargon, and concentrate on *ideas*; you have to think of mathematics as a friend, not as an enemy. I'm not saying that mathematics is always a joyous romp; but you should be able to enjoy it, at whatever level you operate. Do you enjoy crossword or jigsaw puzzles? Do you like playing draughts, or chess? Are you fascinated by patterns? Do you like working out what "makes things tick"? Then you have the capacity to enjoy mathematical ideas. And, perhaps, if you do enjoy them, you might even become a mathematician.

We could do with more mathematicians. Mathematics is fundamental to our lifestyle. How many people, watching a television programme, realize that without mathematics there would be nothing to watch? Mathematics was a crucial ingredient in the discovery of radio waves. It controls the design of the electronic circuits that process the signals. When the picture on the screen rolls up into a tube and spins off to reveal another picture, the quantity of mathematics that has come to life as computer graphics is staggering.

But that's mathematics at work. What this book is about is the flip side: mathematics at play.

The two are not that far apart. Mathematics is a remarkable sprawling riot of imagination, ranging from pure intellectual curiosity to nuts-and-bolts utility; and it is *all one thing*. The last few years have witnessed a remarkable re-unification of pure and applied mathematics. Topology is opening up entire new areas of dynamics; the geometry of multi-dimensional ellipsoids is currently minting money for AT&T; obscure items such as p-adic groups turn up in the design of efficient telephone networks; and the Cantor set describes how your heart works. Yesterday's intellectual game has become today's corporate cash-flow.

However, what you'll find here is the playful side of mathematics, not the breadwinning one. Some items are old favourites, some are hot off the press. Most chapters include problems for you to solve, with answers at the end; there are things to make and games to play. But there's a more serious intention, too. I'm hoping that at least some of you might be inspired to find out more about the remarkable mental world that lies behind the jokey presentation. The ideas you will encounter all have connections with *real* mathematics – though you might be forgiven if you didn't see through the heavy disguise. "Mother Worm's

Blanket" is a problem in geometric measure theory, and "The Drunken Tennis-Player" is about stochastic processes and Markov chains. "Parity Piece" introduces algebraic topology; "The Autovoracious Ourotorus" leads to coding theory and telecommunications. On the other hand, I can assure you that "Close Encounters of the Fermat Kind" has nothing whatsoever to do with space travel or the motion picture industry.

Or does it? Wait a minute . . .

Ian Stewart
Coventry

1

Mother Worm's Blanket

"Bother!" said Mother Worm.

"Something the matter, dear?"

"It's our sweet little Wermentrude. I know I shouldn't criticize the child, but sometimes – well – her blanket's come off again! She'll be chilled to the bone!"

"Anne-Lida, worms don't *have* bones."

"Well, chilled to her endodermic lining, then, Henry! The problem is that when she goes to sleep, she wriggles around and curls up into almost any position, and the blanket falls off."

"Does she move once she's asleep?"

"No, Henry, she sleeps like a log."

She even looks like a log, thought Henry Worm, but did not voice the thought. "Then wait until she's asleep before you cover her up, dearest."

"Yes, Henry, I've thought of that. But there is another problem."

"Tell me, my pet."

"What shape should the blanket be?"

It took a while for Henry to sort that one out. It turned out that Mother Worm wanted to make a blanket which would completely cover her worm-child, no matter how she curled up. Just the worm, you understand: not the area she surrounds. The blanket can have holes. But, being thrifty, Mother Worm wished the blanket to have as small an area as possible.

"Ah," said Father Worm, who – as you will have noticed – is something of a pedant. "We may choose units so that the length of the little horr- . . . dear little Wermentrude is 1 unit. You're asking what shape is the plane set of minimal area that will cover *any* plane curve of length 1. And no doubt you also wish to know what this minimal area *is*."

"Precisely, Henry."

"Hmmmmmmmmm. *Tri*-cky . . ."

When you start thinking about Mother Worm's blanket, the greatest difficulty is to get any kind of grip. The problem tends to squirm away from you. But as Henry explained to his wife – in order to distract her attention from his inability to answer the question – there are some general principles that can form the basis of an attack. Suppose that we know where some points of the worm are: what can we say about the rest? He pointed out two such principles (box 1.1): they depend upon the fact that the shortest distance between two points is the straight line joining them.

"Excellent," said Father Worm. "Now, Anne-Lida my dear, we can make some progress. An application of the Circle Principle shows that a circle of diameter 2 will certainly keep Wermentrude warm. Lay the centre of the blanket over Baby's tail, my dear: the rest of her cannot be more than her total length away! How big is the blanket? Well, a circle of diameter 2 has an area of π, which you'll recall is approximately 3.14159... ."

"That's enough, Henry! I've already thought of something *much* better. Suppose that you (mentally!) chop Wermentrude into two at her mid-point. Each half lies inside a circle of radius $\frac{1}{2}$ centred on her mid-point. If I place a circular blanket of radius $\frac{1}{2}$ – that is, diameter 1 – so that its centre lies over Baby's mid-point, it will cover the dear little thing."

What's the area now? Remember pi-r-squared?

In fact this is the smallest *circle* that will always cover Baby, because if she stretches out straight she can poke out of any circle of diameter less

Box 1.1   Blanket Regulations

The Circle Principle   Suppose we have a portion of worm, of length L, and we know that one end of it is at a point P. Then that portion lies inside a circle of radius L, centre P. The reason: every point on the portion is distance L or less away from P, *measured along the worm*. The straight line distance is therefore also L or less. But such points lie inside the circle of radius L.

The Ellipse Principle   Suppose we have a portion of worm, of length L, and we know where both ends are. Let the ends be at points P and Q in the plane. Form a curve as follows. Tie a string of length L between P and Q, insert the point of a pencil, and stretch the line taut. As the pencil moves, it describes an *ellipse* whose foci are P and Q. The points inside this ellipse are those points X for which $PX + XQ$ is less than or equal to L. Therefore every point on the portion of worm concerned lies inside this ellipse (figure 1.1).

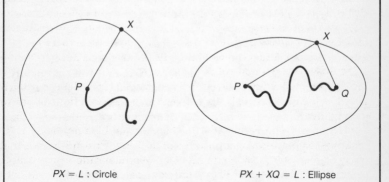

$PX = L$: Circle $PX + XQ = L$: Ellipse

1.1   A portion of worm of length L, *one point* P *of which is known, lies inside a circle of radius* L, *centre* P. *If two points* P *and* Q *are known then the portion lies inside an ellipse with* P *and* Q *as foci, consisting of all points* X *such that* PX + XQ = L.

than 1. But might a shape different from a circle be more economical? "It had better be," groaned Father Worm, who would have to pay for the blanket, as he retired to his study. Two hours later he emerged with several sheets of paper and announced that Anne-Lida's proposal, a circle of diameter 1, is at least twice as large as is necessary.

"Good news, my dear. A *semicircle* of diameter 1 is big enough to cover Baby no matter how much the little pest – er, *pet* – squirms before snoozing."

That cuts the area down even more: to what?

As I said, Henry Worm is a pedant. He won't say anything like that unless he's absolutely certain it's true. So he hasn't just spent his time doing experiments with semicircles: he has a *proof* that the unit semicircle (a semicircle of diameter 1) *always* works. It isn't an easy proof, and if you want to skip it I wouldn't blame you. But proof is the essence of mathematics, and you may be interested to see Father Worm's line of reasoning. If so, it's in box 1.2.

"Very clever, Henry," sniffed Anne-Lida. "But I *think* the same idea shows that you can cut some extra pieces off the semicircle. You see, when P and Q are closer together than b, the distance between X and Y is less than 1. That must leave room for improvement, surely?"

"Hrrrumph. You may well be right, my dear. But it gets very complicated to work out what happens next." And Henry rapidly changed the topic of conversation. My more persistent readers may wish to pursue the matter, because Anne-Lida is right: the unit semicircle is *not* the best possible shape. Indeed, *nobody knows what shape Baby Worm's blanket should be.* The problem is wide open. Remember, it must cover her no matter what shape she squirms into; and you must give a *proof* that this is the case! If you can improve on $\frac{\pi}{8}$, let me know.

Later that evening, Henry suddenly threw down his newspaper, knocking over a glass of Pupa-Cola and soaking the full-size picture of Maggot Thatcher on the front page. "Anne-Lida! We've forgotten to ask whether a solution exists at all!" You can't keep a good pedant down. But he has a point. Plane sets can be a lot more complicated than traditional things like circles and polygons. The blanket may not be convex: in fact it might have holes! For that matter, what do we mean by the "area" of a complicated plane set?

"My God," said Henry. "Perhaps the minimal area is *zero*!"

"Don't be silly, dear. Then there would be no blanket at all!"

Henry poured a replacement and sipped at it with a superior smirk. "Anne-Lida, it is obviously time I told you about the Cantor set."

"What have those horrible horsey snobs got to do with . . ."

"Can*t*or, my dear, not can*t*er. Georg Cantor was a German mathematician who invented a very curious set in about 1883. Actually, it was known to the Englishman Henry Smith in 1875 – but 'Smith set' wouldn't sound very impressive, would it? To get a Cantor set you start with a line segment of length 1, and remove its middle third. Now

Box 1.2 Father Worm's Proof.

A line that meets Wermentrude at some point or points, but such that she lies entirely on one side of it, is called a *support line* (figure 1.2). Support lines exist in any direction. Just start with a line pointing in that direction and slide it until it first hits the worm. Notice that support lines *may* meet the worm in more than one point.

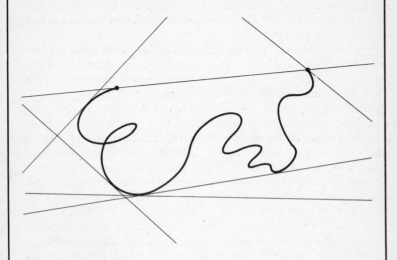

1.2 Support lines.

First, suppose that every support line meets Wermentrude in exactly one point. Then she must be curled up in a closed convex loop, perhaps with other bits of her inside (figure 1.3). Suppose she touches a support line at a point P. Then all points on the loop are at a distance $\frac{1}{2}$ or less from P, measured along the worm; hence also measured in a straight line. So are the other points inside the loop. Therefore Wermentrude lies inside the circle of radius $\frac{1}{2}$ centre P. But she also lies on one side of the diameter of this circle formed by the support line. Thus she lies inside a unit semicircle.

Alternatively, some support line meets Wermentrude in at least two points P and Q. These points divide her into three segments A, B, C of lengths a, b, c, where $a+b+c = 1$ (figure 1.4). The distance between P and Q is at most b because segment B joins P to Q. By the Circle Principle, segment A lies inside a circle centre P radius a; but it also lies on one side of the

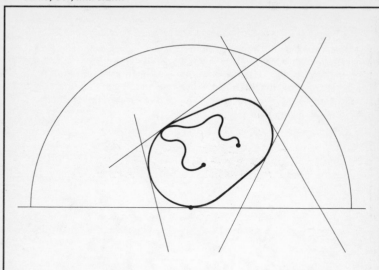

1.3   *If every support line meets the worm in a single point, then the worm determines a convex loop of perimeter less than or equal to 1 and hence lies inside a unit semicircle.*

support line, so it actually lies inside a semicircle of radius *a*. Similarly segment *C* lies inside a semicircle of radius *c*.

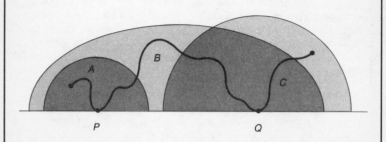

1.4   *If a support line meets the worm in two points, then the worm lies inside a figure obtained by overlapping two semicircles and a semiellipse.*

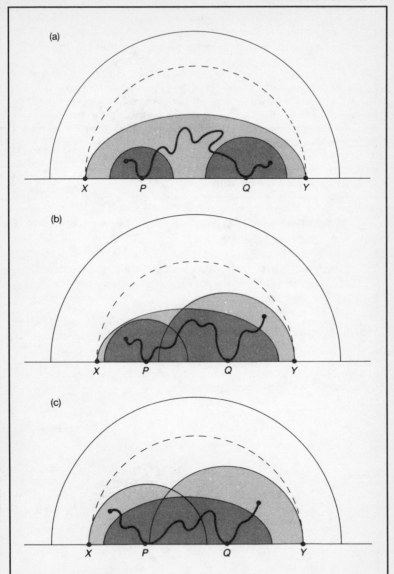

1.5   *There are three possible arrangements of the semicircles and the semiellipse. In all three cases the distance XY is at most 1. Therefore the semicircle on XY as diameter (dotted) fits inside the outer unit semicircle. So does the worm.*

What about segment B? By the Ellipse Principle, it lies within an ellipse whose foci are at P and Q, traced by stretching a string of length b. Because of the support line, B actually lies inside a semiellipse (half an ellipse).

Thus the entire worm lies inside a rather complicated figure formed by overlapping two semicircles and a semiellipse. Let X and Y be the extreme points of this figure on the support line. There is a minor complication now. The point X may be either on the semicircle centre P or the semiellipse; similarly Y may be either on the semicircle centre Q or the semiellipse. However, in each case it is not hard to show that the distance between X and Y is 1 or less (figure 1.5).

Now, the ellipse is "flatter" than a circle; so both semicircles and the semiellipse fit inside a semicircle whose diameter is XY. Since XY is 1 or less, Wermentrude fits inside a unit semicircle.

remove the middle third of each remaining piece. Repeat, forever. What is left is the Cantor set." (Figure 1.6)

"I don't see how there can be *anything* left, Henry."

"Oh, but there is. All the end-points of all the smaller segments are left, for a start. And many others. But you are right in one way, my dear. What is the length of the Cantor set?"

"Its ends are distance 1 apart, Henry."

"No, I meant the length not counting the gaps."

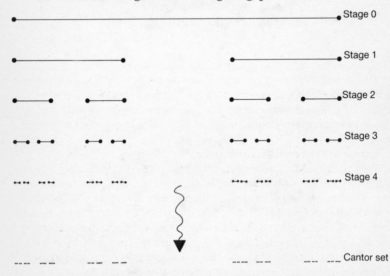

1.6 *Construction of the Cantor set by removing middle thirds. Its length is zero, but it contains infinitely many points.*

"I have no idea, Henry. But it looks very small to me. The set is mostly holes."

"Yes, like Wermentrude's sock."

"Are you criticizing me? I'm going to darn her sock tomorrow! Of all the . . ."

"No, no, my dear; nothing was further from my mind. Hrrumph. The length reduces to $\frac{2}{3}$ the size at each stage, so the total length after the nth stage is $(\frac{2}{3})^n$. As n tends to infinity, this tends to 0. The length of the Cantor set is zero." Anne-Lida worked out the first few powers of $\frac{2}{3}$ on her calculator – it wasn't a pocket calculator because worms don't have pockets – and nodded in agreement.

"Now the Cantor set, despite being mostly holes, has a remarkable property," Henry continued relentlessly. "Given any number between 0 and 1, there are two points in the Cantor set whose distance apart is exactly equal to that number. Er – I don't think you'd want to see a proof, my dear, so let us merely assume the result is true, yes? Good. Now, suppose that Baby can only curl up into rectangles . . ."

"Henry, you know very well she's as wriggly as a baby worm . . ."

"Pretend she's been playing tailball and is very stiff in the joints." Wermentrude, I must add, goes to a non-sexist equal opportunity awormative action school which discourages differences between boy and girl worms – not that you'd notice – and girl worms play tailball just like the boys. Nevertheless, Anne-Lida objected.

"You know very well worms don't *have* joints, Henry!"

"Oh, for heaven's sake! Pretend that they *do*, all right? Just to please me!"

"Very well," said Anne-Lida huffily. "Since you insist."

"Thank you. Because Wermentrude's length is 1, the height and the width of the rectangle are between 0 and 1. So I can find two points in the Cantor set whose distance apart is equal to the height, and two more whose distance apart is equal to the width. Now I consider the *Cantor tartan*."

"Cantor isn't a Scottish name, Henry!"

"Very well, the MacCantor tartan. I take a set of horizontal lines of unit length, spaced vertically according to the Cantor set, together with a set of vertical lines, spaced horizontally the same way (figure 1.7). Now, in the horizontal set I can find two lines whose distance apart is equal to the height of the rectangle, and in the vertical set two lines whose distance apart is equal to its width. So – as J. R. Kinney noticed in 1968 – the MacCantor tartan can be placed so that it covers Baby Worm's rectangle."

"You mean the perimeter, not the inside of the rectangle."

"Naturally. The blanket must cover *Wermentrude*, not the space she encloses."

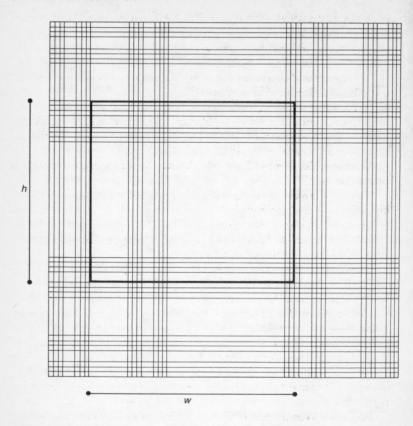

1.7   The MacCantor tartan can cover any rectangle, because the Cantor set contains points distance h *and* w *apart for all* h *and* w.

"That's not a blanket, Henry: it's a net."

"If you wish, I shall rename this chapter 'Baby Eel's Net'. But then your name won't be mentioned . . ."

"No, no, Henry. I now realize it is a cellular blanket."

"Excellent. It also has area *zero*. The horizontal part has area $0 \times 1 = 0$, and so does the vertical part, because the Cantor set has length 0."

"So for rectangular worms," said Anne-Lida, "there exists a blanket of area zero that will cover them all! What a bizarre result!" She paused. "But of course that's because rectangles are very special."

"Well, yes and no," said Henry Worm. "I've been reading about the problem, and it turns out that in 1970 D. J. Ward constructed a blanket of area zero capable of covering any polygonal worm. A worm made up of finitely many straight line segments, that is. The blanket is an incredibly messy tangle, of course – mostly holes."

"Curiouser and curiouser, my dear. And what of smooth worms, like our lithe and flexible Wermentrude?"

"Well, for a while mathematicians began to wonder whether there might exist a zero-area universal blanket for smooth worms – speaking in the vermicular, of course. But in 1979 J. M. Marstrand proved that no blanket of area zero can cover all smooth worms."

"Remarkable. It must have taken some very difficult geometry to prove *that*."

"I gather the main idea was to use the concept of the entropy of a totally bounded metric space, my pet."

"*Fascinating*, Henry! *Do* tell me more."

"Well – hrrrumph – I don't think you'd really want to know about that, Anne-Lida. Ergodic theory is kind of tricky . . ."

"You don't know, do you Henry?"

"Well . . . No. But at any rate, we know that the minimal area for Baby's blanket must be greater than zero."

Mother Worm can be a pedant too. "*Do* we, though, Henry? I mean, might there not be a blanket of area $\frac{1}{2}$ that works, and one of area $\frac{1}{4}$, and one of area $\frac{1}{8}$, and so on – areas greater than zero but becoming as small as we please? Then the minimum area would be zero, but it wouldn't actually correspond to a blanket." *Can you think of a simple problem about minimal areas for which this kind of thing happens? Here's a hint: Mother Gnat's tent.*

But Father Worm knew when he was beaten, and was already talking about the analogous problem in three dimensions: Baby Worm's sleeping-bag. What is the minimal volume that will hold a worm of length 1 in ordinary three-dimensional space? And *that* problem is virtually unexplored. Can you make any progress worming your way towards a solution?

ANSWERS

The circle of radius $r = \frac{1}{2}$ has area $\pi r^2 = \pi (\frac{1}{2})^2 = \frac{\pi}{4}$, which is about 0.785. Easy! Yes, but this is just the worm-up problem . . .

Halving that to get a semicircle leads to $\frac{\pi}{8}$, or about 0.393.

Here's an example of an area-minimizing problem which has solutions of arbitrarily small non-zero area, but does *not* have a solution with area zero. Mother Gnat is making a tent so that her daughter Gnathalie can go

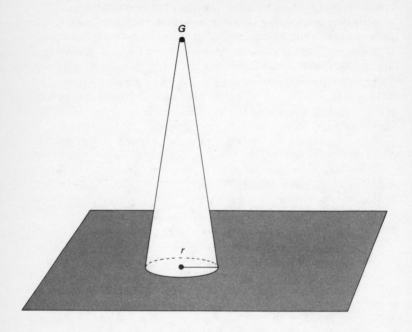

1.8 Mother Gnat's tent, a problem with no minimal solution.

camping. Gnathalie is tiny, no more than a point; she always sleeps hovering a little way off the ground. The tent must cover her head to keep the rain off and reach down to the ground to keep out draughts. What is the smallest area of tent that will do the job?

The answer is that any area greater than zero will work, but *zero itself will not*.

To see why, imagine a point gnat G, some distance – which we may take to be 1 unit – above a plane. The problem of Mother Gnat's tent boils down to this: what is the smallest area of a surface whose boundary lies in the plane, and which passes through G? Consider a sharp conical surface (figure 1.8) whose base is a circle of radius r units. Then the surface area of the cone is πr, and we can make this as small as we want by choosing r small enough. For instance if $r = 0.000000001$ then the area is $\pi r = 0.00000000314\ldots$.

But to get *zero* area we must let $r = 0$, and then the cone becomes a line segment joining G to the plane. But a line segment isn't a surface!

This example shows that area-minimizing problems may not have solutions: that is, the "minimal" area may not be attainable.

Baby Worm's sleeping-bag: do you want to minimize the surface area or the volume? Your choice! Similar arguments can get you to a hemisphere of radius $r = \frac{1}{2}$, with volume $\frac{2}{3}\pi r^3 = \frac{\pi}{12}$, about 0.262; and surface area $3\pi r^2$ (why?) $= \frac{3}{4}\pi$, about 2.356. But it must be possible to improve on those figures.

FURTHER READING

K. J. Falconer, *The Geometry of Fractal Sets* (Cambridge: Cambridge University Press, 1985)

J. M. Marstrand, "Packing Smooth Curves in R^q", *Mathematika*, 26 (1979), pp. 1-12

Herbert Meschowski, *Unsolved and Unsolvable Problems in Geometry* (Budapest: Ungar, 1966)

C. Stanley Ogilvy, *Tomorrow's Math* (Oxford: Oxford University Press, 1972)

D. J. Ward, "A Set of Zero Plane Measure Containing All Finite Polygonal Arcs", *Canadian Journal of Mathematics*, 22 (1970), pp. 815–21

2

The Drunken Tennis-Player

The tennis season has started up again.

A few weeks ago, I spent the afternoon at the local tennis-club, playing an enjoyable match with my friend Dennis Racket. He won in straight sets, 6–3, 6–1, 6–2. Afterwards, as we sank a few beers in the bar, a thought struck me.

"Dennis: how come you always beat me?"

"I'm better than you, old son."

"Yes, but you're not *that* much better. I've been keeping score and I reckon that I win one-third of the points. But I don't win one-third of the *matches!*"

"Let's face it, you don't win *any* matches against *me.*" He took a quick swig at his beer. "That's because you don't win the crucial points, the ones that really matter. I mean, remember when you were leading 40–30 with the set at three games to two? You could have levelled the score at three all. Instead, you . . ."

"Served a double fault. Yes, Dennis, I know all about that. But I reckon I still win about one in three of the *crucial* points! No, there must be another explanation."

"I'd like another *beer*, that's for sure," said Dennis. "My round. I'll be right back." He heaved himself to his feet and began to negotiate his way through the crowd towards the bar. I heard him shouting over the hubbub. "Elsie! Two pints of Samuel Smith's and a packet of peanuts!" With a glass in each hand, he began to make his way back. There were so many people that he went two steps sideways for every step forwards.

Then it hit me.

That's why Dennis always wins!

He sat down, and I decided to share my sudden insight. "Dennis, I've worked it out! Why you always win! I was watching you coming back from the bar, and I suddenly thought: *drunkard's walk!*"

"Actually, my son, they *stagger*. Anyway, I've only had two pints!"

I hastened to reassure him that my choice of phrase was nothing personal. The drunkard's walk – less colourfully called the random walk – is a mathematical concept: the motion of a point which moves along a line, going either left or right, at random. Or on a square grid, taking steps randomly north, south, east, or west. In 1960 Frederik Pohl wrote a science fiction story called *Drunkard's Walk*, and he described it like this:

> Cornut remembered the concept with clarity and affection. He had been a second year student, and their house-master was old Wayne; the audio-visual had been a marionette drunkard, lurching away from a doll-sized lamp-post with random drunken steps in random drunken directions.

To simulate the simplest random walk, all you need is a 30 cm ruler and two coins. One coin acts as a marker, the other as a random number generator. Place the marker coin on the ruler at 15 cm. Toss the other one. If it comes down "heads", move the marker coin 1 cm to the right; if "tails", move it left (figure 2. 1).

According to probability theory, after n moves you will be on average a distance $\sqrt{n}$ cm away from the middle. (Try it!) Despite this, your chances of eventually returning to the middle are 1 (certainty). On the other hand, on average it takes infinitely long to get there. Random walks are subtle things. With a random walk on a square grid, you still have probability 1 of returning to the centre; but in three dimensions the probability of getting back to the centre is about 0.35. A drunkard lost in a desert will eventually reach the oasis; but an inebriated astronaut lost in space has roughly a one in three chance of getting home. Maybe they should have told ET that.

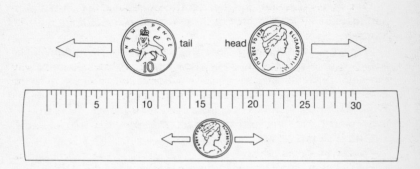

2.1 Apparatus for a random walk.

Years ago a probability theorist told me that the lowest dimensional space in which the chances of getting home are less than 1 is a space of $2\frac{1}{2}$ dimensions, but I've never quite worked out what he meant by that.

As you can see, mathematicians have done a lot of work on random walks. They're important. For example, they model the diffusion of molecules under random collisions in gases and liquids. And they can be used to analyse games of chance.

Such as tennis.

Dennis said he couldn't see the connection.

"But there is one," I said. "Lend me your ears and I'll try to explain why. Let's start with something simpler. Suppose Angus and Bathsheba take it in turns to toss a coin. If it comes up heads, Angus gets one point. Tails, and Bathsheba gets the point instead. Angus wins if he gets three points ahead of Bathsheba; and Bathsheba wins if she gets three points ahead of Angus. If neither has won after ten tosses, the game is a draw. Got that?"

"It's not exactly physically or intellectually challenging, this game," he muttered into his beer.

"Right then, genius: *what is Angus's chance of winning?*"

"Fifty–fifty? Oh, no, they can draw, too. One chance in three."

"I see. He can either win, draw, or lose: you think each is equally likely. Just like tossing a coin: it can either land heads, tails, or on edge, so the chance of it landing on edge is one in three."

Dennis didn't like my sarcastic tone. "All right, cleverclogs: what *is* his chance of winning?"

"I don't know," I said.

"Ha!"

"But if you'll pass me that napkin I'll work it out." And I started to draw a diagram (figure 2. 2).

"What's that?"

"I'm marking Angus's total score along the top, starting at 0, and Bathsheba's down the side. Then I'm going to work out how many ways

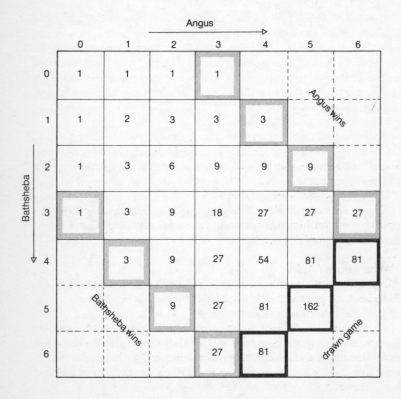

2.2   *Angus and Bathsheba toss a coin.*

the game can reach each legal position. Then I'll count up how many of them are wins for Angus. Well, that's the gist of it, but actually I'll have to be more careful: I'll come to that in a moment."

I wrote a line of 1's along the top and down the side.

"Why all the 1's?"

"They mean – for instance – that there's only one way for Angus to go 3:0 up. He has to win all of the first three tosses."

"Ah."

"But there are *two* ways to get to a score of 1:1."

"I see that. Either Angus or Bathsheba wins the first toss, but then they lose the second."

"Exactly. In other words, the score on the previous turn is either 1:0 or 0:1 in Angus's or Bathsheba's favour – corresponding to the squares above and to the left of the 1:1 square. Each of those contains a 1, and we just add the two numbers up.

"The same method lets us work out how many ways the game can reach any given position, say m:n. The previous position was either $(m-1) : n$ or $m : (n-1)$, and those are the positions above and to the left. Add them up, and write it in. Of course you have to work systematically through the possible scores. For instance, the only reason I know I can put 9 in the 3:2 square is that I've already got 3 in the 3:1 and 6 in the 2:2 positions, OK?"

"Got you."

"And you don't include squares where one player has already won, because the game stops on those. The number at 3:5, for instance, is *not* the sum of the numbers at 3:4 and 2:5, because at 2:5 Bathsheba has won and the game stops."

"It's getting complicated, old lad."

"Nonsense, you just have to be systematic and take the rules of the game into account. Now, Angus wins if the score is 3:0, 4:1, 5:2, or 6:3, and Bathsheba wins for 0:3, 1:4, 2:5, or 3:6. I'll mark those boxes with a shaded border."

"What about 7:4?"

"I said the game stops after ten tosses. That happens at scores of 4:6, 5:5, and 6:4. I'll put a heavy black border on them. *There!*"

We contemplated the diagram.

"Angus wins in 1+3+9+27 ways," said Dennis. "That's 40. He loses in 40 ways, and the game is drawn in 324. That makes 40+40+324 = 404 possibilities altogether. So his chance of winning is $\frac{40}{404}$, which is 0.0990099. About one chance in ten. That sounds unlikely to me, you must have made a mistake."

"Not quite," I said. "*You're* making a mistake. The same one as before: you're assuming each case is equally likely. But because the games go on for different numbers of turns, they *aren't* equally likely."

I bought two more beers and while we consumed them I pointed out that Probability Theory is founded on two bashic prinshiples.

1 To get the probability of a set of distinct events you add the individual probabilities.
2 To get the probability of two independent events happening in turn you multiply their probabilities together.

For instance, if you throw a fair die then the probability of each score in the range 1 to 6 is $\frac{1}{6}$, because all scores are equally likely. The probability of throwing *either* a 5 *or* a 6 is $(\frac{1}{6}) + (\frac{1}{6}) = \frac{1}{3}$. On the other hand, if you throw two dice, say a red one and a blue one, then the probability that the red one is 5 *and* the blue one is 6 is $(\frac{1}{6}) \times (\frac{1}{6}) = \frac{1}{36}$.

"To get the right answer," I told Dennis, "you just apply the rules. At each throw, Angus has a probability $\frac{1}{2}$ of winning, and so does Bathsheba. So each move one square across or down the diagram multiplies the probabilities by $\frac{1}{2}$. The chances of Angus winning 3:0 are $(\frac{1}{2}) \times (\frac{1}{2}) \times (\frac{1}{2})$, or $\frac{1}{8}$. The chances of him winning 4:1 are not $\frac{3}{8}$, but $\frac{3}{32}$, because two more tosses are involved. So his chances of winning are

$$\frac{1}{8} + \frac{3}{32} + \frac{9}{128} + \frac{27}{512}$$

which comes to $\frac{175}{512}$, or roughly 0.3418."

Dennis looked pleased with himself.

"I *told* you he had a one in three chance of winning," he said. Then he added "Ouch!" as I kicked him.

"As a check on the calculation, Dennis, you will observe that the chances of a draw are $\frac{324}{1024}$, the chances of Bathsheba winning are $\frac{175}{512}$, and the sum of the three fractions is

$$\frac{175}{512} + \frac{324}{1024} + \frac{175}{512} = 1,$$

as it must be if I haven't made any mistakes."

"You're a genius. Now, *what's all this got to do with tennis?*"

"It's the same thing, only with different rules. Tennis is a series of *points*, leading to *games*, leading to *sets*, leading to a *match*. To keep it simple, suppose Angus and Bathsheba play one *game* of tennis. On each separate point, Angus either wins or loses; and Bathsheba loses or wins. The winner of the game is the first player to get four points. Unless the score gets to three all, in which case . . ."

"Three all? Three all? What kind of tennis score is *three all*?"

"Deuce. Look, tennis has this incredibly silly scoring system that goes 15, 30, 40, *game* instead of 1, 2, 3, 4, that's all. The '40' is really '45' but people got lazy; I suppose a game is really 60. There must have been a reason originally, but I have no idea what it was and it's just traditional now.

"When the score gets to deuce, the game continues until one or other player gets *two points ahead*.

"You can represent a tennis game on a diagram just like the one we drew for the coin-tossing game." I went over to the bookshelf, came back with a book of tennis scores, and picked one at random. "Look, here's the fifth game of the second set of the 1987 men's singles final at Wimbledon. Pat Cash v. Ivan Lendl. Cash was leading 3–1 in the second set and one set to love. Lendl served and lost. Here's how the scoring went." (Figure 2. 3)

"Oh, I see. It's quite clever, the way a deuce game chases off down that funny zig-zag."

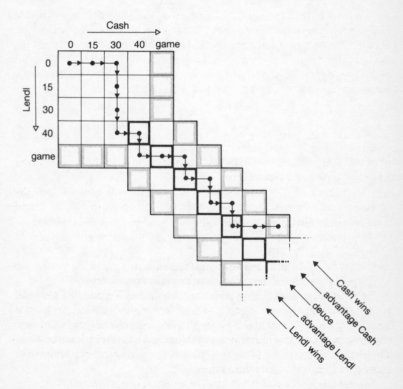

2.3   *Pat Cash and Ivan Lendl play tennis: Lendl to serve.*

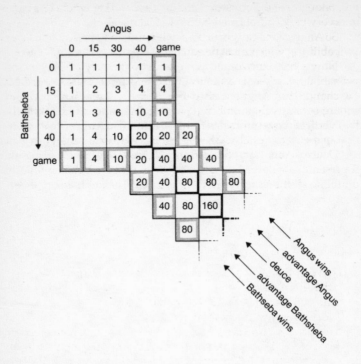

2.4   Combinatorics of a game of tennis.

"The same thing comes up in tie-breaks too: you'll see when I get that far. In principle a game can go on forever. Of course, the chances of that are zero.

"Anyway, you can assign numbers to the squares in just the same way: each square contains the sum of the numbers immediately above and to the left, unless those squares represent the end of a game, which only matters in a deuce game (figure 2. 4).

"Of course, the chances of a given player winning a point aren't $\frac{1}{2}$ any more. Better players win points more often, as I was complaining when we first sat down. Now, to keep it simple, I'm going to assume that the chance is always the same on each point."

Dennis started to object. "But . . ."

"But players have more chance of winning if it's their serve – yes, I know that. Let's keep it simple to start with, though. The same method will take care of different probabilities depending who's serving, but it gets very very complicated if you do that.

"So Angus wins any given point with probability p, say, and loses with probability q, which must be equal to $1 - p$.

"Now every horizontal move is a point won by Angus, so has probability p, whereas a vertical move has probability q. For instance, the chance that Angus wins game–30 is $10p^4q^2$ because the game–30 square contains the number 10, and is four squares horizontally and two vertically away from the starting-point. His total chance of winning is $p^4 + 4p^4q + 10p^4q^2$, plus whatever happens when the game goes to deuce.

"Deuce scores complicate things a bit. But see how the numbers representing wins for Angus run down the diagonal: $10, 20, 40, 80, 160, \ldots$ doubling all the time. We have to add up an infinite series

$$10p^4q^2 + 20p^5q^3 + 40p^6q^4 + 80p^7q^5 + \ldots$$

and then add on $p^4 + 4p^4q$. Now the infinite series is

$$10p^4q^2(1 + 2pq + 4p^2q^2 + 8p^3q^3 + \ldots)$$

and the expression in brackets is a *geometric progression*."

"I did those at school!"

"Can you remember what the sum is?"

"No. Never saw much point to that stuff."

"$1 + r + r^2 + r^3 + \ldots = \frac{1}{(1-r)}$. Provided $-1 < r < 1$, of course. *Now* you see how useful it is! Frankly, I'm amazed you play tennis so well, not knowing how to sum a geometric progression. Anyway . . . Each term is $2pq$ times the previous one, so the expression in brackets is $\frac{1}{(1-2pq)}$. That makes Angus's chance of winning exactly

$$p^4 + 4p^4q + \frac{10p^4q^2}{1-2pq}$$

Isn't that beautiful!"

"Beauty," said Dennis, "is in the eye of the beholder. Let me buy you another beer. You must be thirsty after all those calculations." He wobbled to his feet. "I know *I* am," he muttered, as he took a tentative step forward.

While he was fighting his way back to the bar, I worked out my chances of winning a game against him, assuming my chance of winning a point was one in three. That made $p = \frac{1}{3}$, $q = \frac{2}{3}$, and the formula gave me a probability of $\frac{35}{243} = 0.144$. About $\frac{1}{7}$.

"Dennis: if I have a one in three chance of winning each point, I only have a one in seven chance of winning a game! No wonder you always beat me! The rules of tennis amplify differences between players. I bet the ampifli- . . . amflipi- . . . I bet it gets even bigger when you take sets and matches into account!"

"Very likely, old son. But it's time to go home."

"Why? I was just getting . . ."

"The bar's closing."

To ease my hangover I spent the next morning working out what happens when you take sets and matches into account. The methods are just the same as those I've described, so I'll just summarize the results.

First, let's recall the rules.

In men's singles, a match consists of a maximum of five sets. A player must win at least three sets, and be two sets or more ahead of his opponent, except for a score of 3–2.

To win a set, a player must win at least six games, and be two or more games ahead. A set in the position 6–5 or 5–6 continues for a further game, and is won if the score goes to 7–5 or 5–7. If a set reaches 6–6 it proceeds to a *tie-break*, except for the fifth set in a match, in which case it continues indefinitely until one player is two games ahead.

A tie-break is much like a normal game. However, the scoring goes 0, 1, 2, …, like the games in a set rather than the points in a game. To win, you must score at least 7, and be at least two points ahead.

Before the tie-break rule was introduced, all sets continued until one side was two games ahead. In a doubles match on 15 May 1949 F. R. Schroeder and R. Falkenburg played R. A. Gonzalez and H. W. Stewart (all of the USA) and won the first set by the margin of thirty-six games to thirty-four! The final score was 36–34, 3–6, 4–6, 6–4, 19–17, and the match took four and three-quarter hours.

You can see why the rules were changed.

The diagrams for a tie-break game, a set with or without a tie-break, and a match, are shown in figures 2.5–2.8. The corresponding formulas for probabilities of winning are shown in box 2.1. You should be able to see how they are derived from the diagrams. Capital P means "probability of winning" whatever follows it in brackets. If the play can continue indefinitely, the formula includes the sum of an infinite geometric progression.

The rules for women's singles are slightly different. A match can be won either two sets to love or two sets to one. Tie-breaks are played in every set. You might like to carry out this analysis yourself.

By fitting all the formulas together you can, in principle, write down an explicit expression for the probability of winning a tennis-match. I've indicated with arrows in box 2.1 how to do this: substitute for p the expression in the box at the tail of the arrow, and one minus this for q. I haven't actually carried this procedure out, because the result would be *enormous*. Each single p or q in one formula becomes an entire expression from the previous formula, and the complications become horrendous.

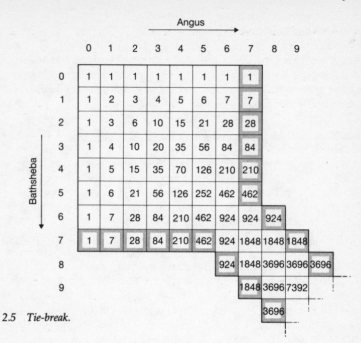

2.5 *Tie-break.*

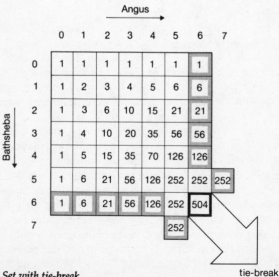

2.6 *Set with tie-break.*

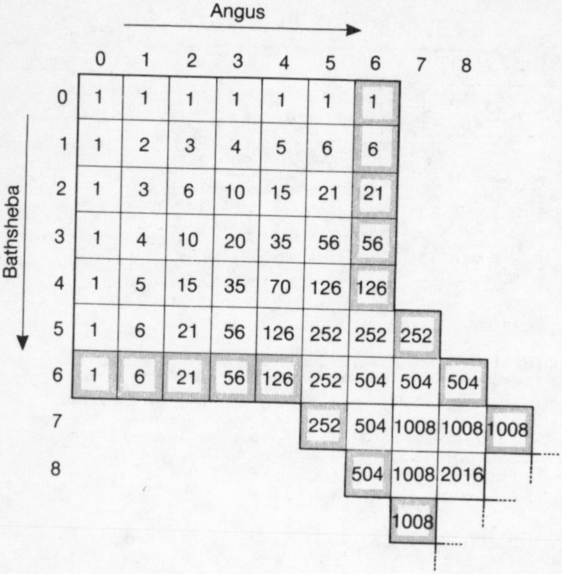

	0	1	2	3	4	5	6	7	8
0	1	1	1	1	1	1	1		
1	1	2	3	4	5	6	6		
2	1	3	6	10	15	21	21		
3	1	4	10	20	35	56	56		
4	1	5	15	35	70	126	126		
5	1	6	21	56	126	252	252	252	
6	1	6	21	56	126	252	504	504	504
7						252	504	1008	1008	1008
8							504	1008	2016
							1008		

(Angus along the top, 0–8; Bathsheba down the side, 0–8)

2.7 *Set without tie-break.*

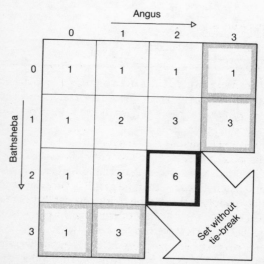

2.8 *Match.*

Box 2.1 Game, set, and match: probabilities of winning

Game

$p = P(\text{point}), q = 1 - p$

$$p^4 + 4p^4q + \frac{10p^4q^2}{1-2pq}$$

Tie-break

$p = P(\text{point}), q = 1 - p$

$$p^7 + 7p^7q + 28p^7q^2 + 84p^7q^3 + 210p^7q^4 + \frac{462p^7q^5}{1-2pq}$$

Set with tie-break

$p = P(\text{game}), q = 1 - p$

$$p^6 + 6p^6q + 21p^6q^2 + 56p^6q^3 + 126p^6q^4 + 252p^7q^5 + 504p^6q^6 P\,(\text{tie-break})$$

Set without tie-break

$p = P(\text{game}), q = 1 - p$

$$p^6 + 6p^6q + 21p^6q^2 + 56p^6q^3 + \frac{126p^6q^4}{1-2pq}$$

Match

$p = P(\text{set with tie-break}), q = 1 - p$

$$p^3 + 3p^3q + 6p^2q^2 P(\text{set without tie-break})$$

However, you can substitute values from one formula to the next, and I've shown what happens in figure 2.9. This gives a table, and a graph, of the probability of winning a men's singles match if your probability of winning any individual point is p.

I showed all this to Dennis the next evening.

"Should Bathsheba be playing men's singles?" he objected.

"She's very liberated. She's thinking of changing her name to Boris. Now, shut up and listen. Observe that the graph is very flat at each end but rises extremely steeply in the middle. With a probability of more than 0.6 of winning each point, your chance of winning the game is nearly 1. *The rules of tennis favour the better player.*"

He stared at me over his beer, perplexed. "But they should, shouldn't they? I mean, the better player ought to have the better chance of winning."

probability of winning	
point	match
0	0
0·1	0
0·2	10^{-22}
0·3	$4·5 \times 10^{-11}$
0·4	$4·4 \times 10^{-4}$
0·5	0·5
0·6	0·9995
0·7	0·9999
0·8	0·9999
0·9	1
1·0	1

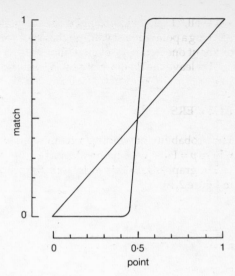

2.9   Calculating the winning probability.

"True."

"But you say all this depends on the assumption that the probability of winning a point is always the same. That's not very realistic."

"You're referring to the advantage of serving."

"Right! When a player is serving, he stands a much better chance of winning the point than when he's receiving – present company excepted, of course."

"Hmph."

"Shows how important the serve is."

"I could redo the calculations . . ."

"Not on my account. I've got the message. You can apply probability theory to tennis." He sank mockingly to his knees and bowed his head to the floor. "I believe, I believe!"

I ignored his antics. "Mmm, but it might be interesting . . . You see, the way the scoring amplifies any advantage means that each player has a chance rather close to 1 of winning his service game – provided his chance of winning a point is above $\frac{1}{2}$. That tends to act the *opposite* way, which evens the game out again! Where's that pencil? . . ."

"Hang on," said Dennis, heaving himself back into his chair. "Before you cover the tablecloth with algebra, answer me one thing. On this theory of yours, what chance do you have of beating *me*?"

"Well," I said, "according to my calculations, if I have a $\frac{1}{3}$ chance of winning a point against you, my chance of winning a match is 0.000000027, or about one in thirty-seven million."

"I'd leave the theory just as it is," he said. "It looks perfect to me."

ANSWERS

The probability of winning a set in women's singles tennis is $p^2 + 2p^2q$, where $p = $ P(set with tie-break) and $q = 1 - p$.

The graph of how this varies with the probability $p = $ P(point) is shown in figure 2.10.

probability of winning	
point	match
0	0
0·1	10^{-29}
0·2	$1·4 \times 10^{-15}$
0·3	$8·4 \times 10^{-8}$
0·4	$3·9 \times 10^{-3}$
0·5	0·5
0·6	0·9961
0·7	0·9999
0·8	0·9999
0·9	1
1·0	1

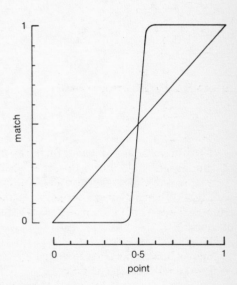

2.10 Calculating the winning probability for women's singles.

FURTHER READING

R. Hersh and R. J. Griego, "Brownian Motion and Potential Theory", *Scientific American* (March 1969), pp. 66–74

Mark Kac, "Probability", *Mathematics in the Modern World*, ed. Morris Kline (San Francisco: Freeman, 1968)

Morris Kline, *Mathematics in Western Culture* (Harmondsworth: Penguin, 1972)

A.N. Kolmogorov, "The Theory of Probability", *Mathematics: its Content, Methods, and Meaning*, ed. A. D. Aleksandrov (Boston: MIT Press, 1963)

Frederik Pohl, *Drunkard's Walk* (London: Gollancz, 1961)

Warren Weaver, *Lady Luck* (New York: Dover, 1963)

3

The Infinormatics Laboratory

The phone rang. It was Philippe Boulanger, the editor of *Pour la Science*. "This month's special topic is Large Computer Systems. I want you to write something about it for 'Visions Mathématiques'."

I protested. "That's 'Computer Recreations'! My column is for people who haven't got computers, don't want computers, maybe even *detest* computers . . ."

"I know you can do it," he said. "Deadline's Thursday week." And he hung up.

I broke into a cold sweat, close to panic. I needed help, and I needed it fast. I had to pick the brains of an expert. A ray of light . . . A quick visit to my old friend Dr Zebedee J. J. Bunnydew, at Salmigondis Corporation's main research centre in the city of Kluzmopodion. That's not as easy as you might think, because Kluzmopodion is on the planet Ombilicus, a billion light-years (and a few metres) from Earth, in the direction of the right eye of the constellation Orion. However, there's a space-time warp in one corner of my garden (behind the raspberry canes), which leads directly to Ombilicus. This is how I met Dr Bunnydew in the first place: I fell through the warp while weeding. Making sure the neighbours weren't watching, I stepped through, and hitched a ride to Salmigondis Corporation on a passing Brontëosaurus-cart.

"Large Computer Systems?" said Bunnydew. "I can tell you a lot about those. Only I can't."

"You can only you can't?"

"Top Secret. Classified government work." He leaned closer. "Quasar Wars contract," he whispered. "It's so secret I can't even talk to myself about it." "But," he added, "you're in luck. I've got some ideas that are so crazy I haven't told the security people about them yet. *Large* computer systems? They don't come any larger than what I'm planning, I can tell you! Follow me!"

He led the way down a corridor to a tiny room. There was a sign pinned to the door, a single symbol: ∞. Perhaps it was room 8, and the sign had slipped. But I didn't think so. I'd seen that symbol somewhere before.

"This," said Bunnydew in a conspiratorial tone, "is the Infinormatics Laboratory."

Infinor- . . . Of course! It was an *infinity* symbol! But what the devil *was* infinormatics? I was soon to find out.

We went in. He opened a drawer and pulled out a length of black plastic, about 5 mm wide. There was a double row of metal pins down the side. I could see about a metre of the thing; then it disappeared into the drawer.

"It looks like an integrated circuit chip," I said. "But longer."

"Much longer," he said. "You are looking at one end of the Bunnydew Infinite Linear RAM chip – a computer memory with infinitely many locations, each capable of storing a single binary digit in electrical form. If electricity is present in a given location, then the digit is a 1, otherwise it is a 0. Sequences of 0's and 1's can code any information whatsoever. One BILRAM can store not just all the information in the universe: it can store an infinite amount of information!"

"I see why you leave most of it in the drawer."

"Well, yes, it is rather unwieldy. I have to store it in an infinite pandimensional compressor field, but don't worry about technical details."

"Doesn't it take electrical signals a long time to get from one end to the other?" I asked.

"Strictly speaking it only has one end, the one I'm holding. The other 'end' just goes on forever. But yes, it takes infinitely long."

I suggested that this wasn't very practical. Bunnydew agreed. "But the BILRAM is still very interesting. It needs no power source."

I didn't believe him. "What about the Law of Conservation of Energy?"

"Doesn't apply," he said airily. "Not to an infinite system. Let me explain. The BILRAM is made from silicon, which is a semiconductor. Its memory locations work electrically. Electrical power is obtained from electrons. Now, if you remove an electron from a semiconductor, you get what physicists call a *hole*.

"Suppose that I start with a BILRAM in which every memory location contains a binary zero: no electrons. Also no holes: neutral. Understand?"

"Sure."

"Good. Now, I create an electron in location 1 by borrowing it from location 2."

"But that leaves a hole in location 2! Energy is conserved!"

"You speak too soon, my boy. Because I also remove an electron from location 3 and place it in location 2. That fills up the hole in location 2, but of course creates one in location 3. I get rid of that by borrowing an electron from location 4. Suppose I borrow electrons infinitely many times. For every integer $n = 2, 3, \ldots$ I remove an electron from location n and place it in location $n - 1$ (figure 3.1). What do I get?"

I thought about that. "You get an electron in location 1. The other locations all lose one electron but gain another . . . So they stay neutral."

"But," I added, "of course you get a hole at infinity."

"No I don't," he said. "Infinity never comes into it. Each location corresponds to a *finite* value of n. No, I get an electron in location 1 while everything else ends up the same as it started. Creation from nothing! That's just one of the paradoxes of infinity. And that's not all. Imagine a BILRAM filled up completely with an infinite amount of information . . ."

"Huh? How can you have an infinite amount of information?"

"How about a complete list of all prime numbers? Of course you can have an infinite amount of information! I've got a full BILRAM here in the lab – it's called the GALILEO file (figure 3.2). I bet you can't guess what's on it! Anyway, suppose you've got a BILRAM that's completely full, and you want to add another piece of information to it. What do you do?"

"You can't do anything! If the chip's full, there's no more room!"

"On an infinite chip there is. Forget the electrons, and think about the way the *holes* moved." I thought. Suppose you have a list of binary digits which should read 101100011000... but you've forgotten the 1 at the front.

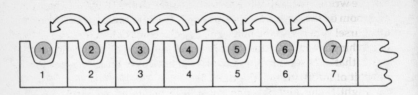

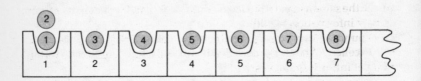

3.1   *Electron-hole pairs are electrically neutral. If electrons (black) are moved through an infinite system of holes, it is possible to create a free electron from nothing, in violation of the Law of Conservation of Energy.*

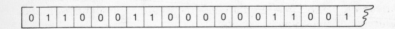

3.2   *The beginning of the* GALILEO *file, containing an infinite amount of information. What does the file list and how is the information coded?*

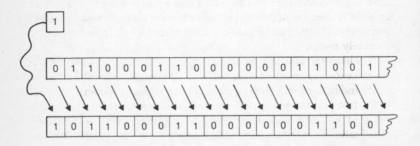

3.3   *To add a digit 1 at the front of the full* GALILEO *file, just move each digit up one place.*

You've written them all into a BILRAM, in order, 01100011000.... There isn't any room on the far end for the missing 1, because there isn't a far end ... Of course! Where did the holes go? The electron trick in reverse! "Move everything *up* one location," I said. "The information in location 1 goes into 2, that in 2 goes into 3, and so on ... That leaves location 1 free for the new bit of information." (Figure 3.3)

"Right. So infinity plus one is just infinity again." He wrote "$\infty + 1 = \infty$" on a notepad. "One of the many paradoxes of the infinite. The whole can be the same as a part. But let us continue. What if we have *several* items of new information to add?"

"You move everything up sufficiently many times."

"Excellent! And this proves that if you add a finite number to infinity, you get infinity again, yes?"

"I suppose so ... It all depends on what you mean by 'add'."

"You are beginning to be cautious about infinity. I like that. No doubt you can now work out how to add an *infinite* amount of new information to a full BILRAM?"

"Move it up an infinite number of ... Oh, no, it all falls off the end at infinity."

"But there isn't an end at infinity."

"It still falls off," I said doggedly. "I'm sure it does. Even if there isn't an end to fall off *from*. If you move the information along an infinite number of times, you lose it all."

"Correct."

"It can't be done, then."

Zebedee J. J. Bunnydew laughed. "So infinity plus infinity makes a *larger* infinity?"

"Yes. No! I'm confused! Infinity is the biggest thing there is. You can't have *two* different sizes of infinity ..."

He shook his head sadly. "Wrong again. Your terrestrial mathematician Georg Cantor would be turning in his grave. But that is irrelevant. To add an infinite amount of information to a full BILRAM, you merely move the contents of location n to location $2n$. That frees up all the odd locations – infinitely many."

"It's like riffle-shuffling a pack of cards!" I said in excitement (figure 3.4).

"An excellent image, my friend. Yes – if you take two infinite packs of cards and riffle them together, you get just one pack, the same size as each of the two you started with. So $\infty + \infty = \infty$, as you might expect.

"And you can even accommodate infinitely many sets of infinite information. You begin to see the attractions of my BILRAM! A memory that never fills up; or rather, if it does, you just move the contents around and create new space from nowhere!"

I pointed out that all this took infinitely long to happen. "You asked

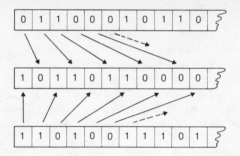

3.4 Riffle-shuffling two infinite lists together creates a single list the same size.

me about *large* computer systems," he replied. "Not *fast* ones." He grinned. "Of course, if I could make the first move in one second, the next in half a second, the next in a quarter of a second, and so on, then after two seconds, the job would be finished!"

"Ridiculous. You can't move faster than the speed of light."

He looked at me archly. "So how did *you* get here, a billion light-years (and a few metres) from Earth, without bringing so much as a sandwich with you to eat on the journey?"

I blushed. "Well, except for space-time warps . . ."

"To name but one method. I'm working on an improved design of infinite RAM which avoids the problem entirely." He opened another drawer and pulled out a sketch (figure 3.5). "The Bunnydew Golden RAM chip," he said proudly. "Let me remind you of the two basic principles of chip manufacture. One: repetition. Two: photographic miniaturization. You will note that the design consists of infinitely many repetitions of the same basic unit, but continually reduced in size. The basic unit, of course, is a single memory location."

Zebedee J. J. Bunnydew had developed a new infinite-zoom camera that could reduce a photograph to any size, however small. His basic memory circuit occupied a square. He had solved the problem of fitting infinitely many squares efficiently on a rectangular chip by using a rectangle whose sides were in the *golden ratio* φ. If a square is removed from such a rectangle, the remaining rectangle is exactly the same shape as the original. The exact value of φ can be calculated from this, and it is $\frac{(1+\sqrt{5})}{2} = 1.618034\ldots$.

"This design," said Bunnydew proudly, "ensures that if one memory location is removed, the remaining part of the chip is exactly the same as the whole thing, but reduced in size. My new recursive photography techniques . . . Oops, they're classified – forget I said anything. Anyway,

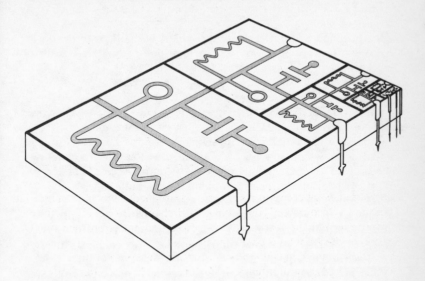

3.5 The Bunnydew Golden RAM chip, in which the same memory unit is repeated infinitely often.

because the total size is finite, the speed of light is no longer a limiting factor. Also, it's the only chip in existence that can be made as small as you like – and therefore as fast as you like – just by cutting bits off."

"That's impressive," I admitted. A thought struck me. "Does it work? What happens when the units get below atomic size?"

"I shrink the atoms too," he said.

"Somebody round here needs a shrink."

"I'll admit," he said, "that I've run into trouble making an infinite-resolution photographic emulsion. But I'll find a way eventually." He put the design back in the drawer and shut it. "And when I do, I'll also be able to realize my design for an infinitely accurate digital watch which shows the time correct to an infinite number of decimal places by using a golden rectangle liquid crystal display which is *almost . . .*"

To calm him down, I tried to distract him. "I know a puzzle about infinite machines," I said. "Imagine a light switch. You switch it on. After one second you switch it off. After a further half second you switch it on again. After another quarter second you switch it off, and so on. On–off–on–off faster and faster, each taking half the time of the previous one. After two seconds, you've switched it on and off infinitely many times. Got that?"

"I understand."

"The question is: after those two seconds, is the light on or off?" Bunnydew's face took on a mysterious look, and he hesitated. After a time, he said: "Off."

"Why?"

"Because you'll blow the fuse if you switch on and off that fast."

I aimed a kick at him, but he dodged. "Hey!" he said, cowering on the far side of a laboratory bench, "You've just given me a *wonderful* idea. Computers are just combinations of switches. I could design a computer like that light switch! BUNNYRAC, the Bunnydew Rapidly Accelerating Computer! Do you realize how powerful a computer would be if you could perform infinitely many computations in a finite time?"

"Well, more powerful than a Cray, for sure . . ."

"Name a famous unsolved problem in mathematics. Any will do."

"Um. Goldbach's conjecture. Every even number is a sum of two primes. Proposed by Christian Goldbach in a letter to Leonhard Euler on 7 June 1742, still unsolved."

"Fine. On BUNNYRAC, you could prove or disprove Goldbach's conjecture by trial and error. In the first second, you try all possible ways to represent 2 as a sum of two primes, getting $2 = 1+1$ of course."

"But 1 isn't a prime."

"It isn't considered to be one now. It was in Goldbach's day. Otherwise his conjecture would have been obviously false. Stop splitting logical hairs! Where was I? Oh, yes . . . In the first second you test the number 2 to see if it is a sum of primes. In the next half second, you test the number 4. In the next quarter second, you test the number 6; in the next eighth of a second you test 8, and so on. After two seconds, you've tested every possible even number! Either you prove the Goldbach conjecture, or you find an example that proves it false. A completely foolproof method."

"Wow!" I was excited now. "You could solve other problems that way, too! Fermat's Last Theorem – a perfect nth power can't be the sum of two other perfect nth powers, for all $n \geq 3$. You just try all n in turn, faster and faster! You could prove or disprove the Riemann hypothesis by calculating all the infinitely many zeros of the zeta function! You could find out whether or not there are infinitely many twin primes – primes that differ by 2, like 19 and 17 – by testing every possible pair! You could . . ."

" . . . get more ambitious. You still haven't grasped how *big* infinity is. With BUNNYRAC you could prove *every possible theorem in mathematics* in two seconds."

"*What?*"

He sighed. "To prove theorems, you start with a small number of basic statements – axioms, things you assume are true – and apply a small number of rules of deduction. That's what a proof is. There are infinitely many possible proofs, hence infinitely many possible theorems; but there are only finitely many theorems whose proof is a given length. That means you can arrange all possible proofs in order, and check them one by one, faster and faster, in a finite time."

"That," I said, "is positively bizarre. Also horrifying. It would put mathematicians out of a job forever!" I knew what Pandora must have felt like when she opened that box . . .

Zebedee J. J. Bunnydew emerged from behind the bench and sat me gently on a stool. "Don't worry, there are a few bugs to iron out yet. How long do you think it will take a human being to *read* the list of all possible theorems if BUNNYRAC prints one out?"

ANSWERS

What does the GALILEO file

0110001100000011001...

contain, and how is the information encoded? The name is a clue, as I'll explain in a moment. The answer is "a list of all squares". The squares are in binary, and the coding is as follows. A sequence of n 0's, terminated by a 1, means that "the next n digits are the next square on the list". Then the square follows; after which another row of 0's terminated by a 1 says how long the next square is. (These sequences 00 ... 01 are needed so that you know where a given entry in the list starts and finishes.) So the file decodes as

01	a 1-digit number follows
1	it's 1
0001	a 3-digit number follows
100	it's 4 (in binary)
00001	a 4-digit number follows
1001	it's 9 (in binary)

and so on.

The name? In Galileo's *Mathematical Discourses and Demonstrations* of 1638 the sagacious Salviati notes that "to each square corresponds its root", that is, there are exactly the same number of perfect squares as there are whole numbers – even though most whole numbers are not squares! Another paradox of infinity.

FURTHER READING

Michael Guillen, *Bridges to Infinity* (London: Rider, 1983)
Edward Kasner and James Newman, *Mathematics and the Imagination* (London: Bell, 1961)
Eugene P. Northrop, *Riddles in Mathematics* (Harmondsworth: Penguin, 1960)
Ian Stewart, *The Problems of Mathematics* (Oxford: Oxford University Press, 1987)
Leo Zippin, *Uses of Infinity* (Washington, DC: Mathematical Association of America, 1962)

4

The Autovoracious Ourotorus

A mythical serpent of ancient Egypt.
An alchemical symbol.
Kekulé's discovery of the benzene ring.
An Indian theory of rhythm, a thousand years old.
The seven bridges of Königsberg.
The theory of telephone circuits.
Radar maps of Venus.

A random assortment of items? Not at all. They all have something in common – but you'll never guess what.

The common thread is a nonsense-word in Sanskrit: *yamátárájabhánasalagám*.

The curious unity of these ideas was discovered in about 1960 by Sherman K. Stein, a mathematician at the University of California, Davis. Much of the tale that I shall tell is based on chapter 8 of his book, *Mathematics: the Man-Made Universe*. But there's a twist to the tale – rather, tail – and you'll find some new material here too.

The mythical serpent of ancient Egypt is the worm Ouroboros, which puts its tail in its mouth and continually devours itself. It was used as an alchemical symbol in the Middle Ages. The chemist Friedrich Kekulé invented his famous ring structure for the benzene molecule after dreaming about Ouroboros. A similar "tail-eating" concept occurs in the musical theories of ancient India, by way of the nonsense-word above. This word raises a mathematical problem which can be solved by applying ideas invented by Leonhard Euler to solve the famous problem of the Königsberg bridges. The results have applications to telephone transmission and the methods used to map the surface of Venus from the Earth using sensitive radar.

It's a curious story.

And it involves some delightful recreational mathematics, which poses many unsolved problems, suitable for the amateur to tackle.

Stein found out about *yamátárájabhánasalagám* from a composer, George Perle, who told him that it was a word invented as a mnemonic for rhythms. What's important is not the vowels and consonants, but the stress placed on the syllables. Perle explained it this way: "As you pronounce the word you sweep out all possible triples of short and long beats. The first three syllables, *ya má tá*, have the rhythm short, long, long. The second through the fourth are *má tá rá*: long, long, long. And so on." There are eight distinct triples of rhythms, long or short; you can check that each occurs in the nonsense word exactly once.

Stein reduced the word to its mathematical content by using 0 for short and 1 for long, so that it became 0111010001. "After staring at the simplified string for a while, I noticed a lovely thing. The first two digits are the same as the last two; if I bent the string into a loop, it would look like a snake swallowing its own tail." (Figure 4.1) He called the result a *memory wheel*, because you can start in any position, and by clicking round one space at a time, generate all possible triples of digits 0 and 1:

```
... 0 1 1 1 0 1 0 0 ...
    0 1 1
      1 1 1
        1 1 0
          1 0 1
            0 1 0
              1 0 0
                0 0 0
                  0 0 1
```

Like that.

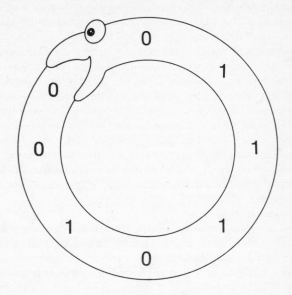

4.1 An ouroborean ring containing all triples of digits 0 and 1.

Let's use a more impressive name, and call it an *ouroborean ring*.

Any alert mathematician immediately starts asking questions. Are there ouroborean rings for quadruples of 0's and 1's? Quintuples? *n*-tuples? What about 0's, 1's, and 2's? Vast generalizations flash through the mathematician's fertile mind.

Then a simpler question occurs. What about *pairs*? Can you find a sequence of length four, consisting of only 0's and 1's, which when written in a circle contains all four possible pairs 00, 01, 10, 11? Try it. When you've solved that one – which is easy – try finding an ouroborean ring containing all sixteen quadruples. Then read on.

Yes, there is an ouroborean ring for pairs: 0011 (figure 4.2). It's essentially unique: all other solutions can be found by rotating it to get 0110, 1100, 1001, which look the same when you draw them on a self-devouring snake.

Stein found one for quadruples:

1111000010100110.

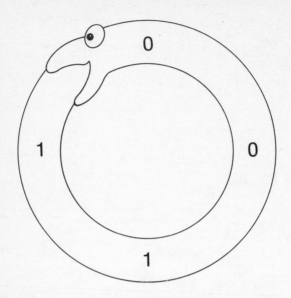

4.2 *The unique ouroborean ring containing all pairs of digits 0 and 1.*

By this time he was convinced that there would be ouroborean rings for *n*-tuples of 0's and 1's. He tried to prove it, but although he had a very clever idea, it led to an unsolved problem in mathematics. But he then discovered that I. J. Good had come across the question in some research in number theory in 1946 – *and had solved it*.

Good's main interest was in finding an endless sequence of 0's and 1's in which every possible sextuplet occurred equally often, but his method was more general. He used a trick to turn the problem into one that had been solved in 1735 by Leonhard Euler, the most prolific mathematician in history. Euler's problem is one of the early sources of topology. Although it's extremely well known, I'll reproduce it here. The subsequent history of the Königsberg road system, also important to our story, is less well known (probably because I've just invented it).

"In the town of Königsberg," wrote Euler, "there is an island called Kneiphof, with two branches of the River Pregel flowing around it. There are seven bridges [figure 4.3(a)]. The question is whether a person can plan a walk in such a way that he will cross each of these bridges once but not more than once."

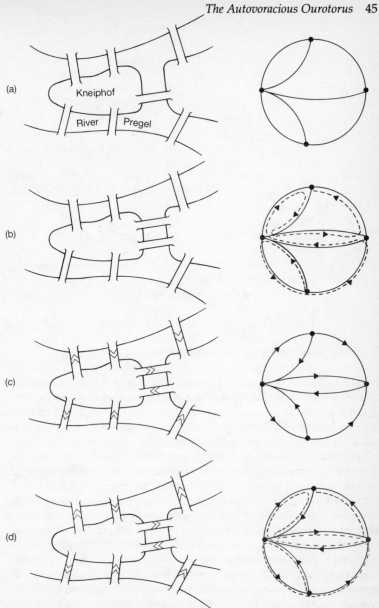

4.3 (a) *Euler's problem of the Königsberg bridges: road map, and corresponding graph.*
(b) *After building the Kneiphof relief road.* (c) *When the one-way system was introduced.*
(d) *How the one-way system was changed so that Euler's theorem applied.*

You may wish to experiment. You'll soon decide that there's no solution.

Euler went further. He *proved* there is no solution, and he found general conditions for the existence of solutions to any problem of the same kind.

To do this, he replaced each land mass by a *dot* and each bridge by a *line* connecting appropriate dots, obtaining a *graph* which accurately reflects the topology of the connections. The graph is shown on the right of the figure, at the top. The problem now becomes: can you trace a path through the graph, passing exactly once along each edge?

Well, said Euler: suppose such a path exists. Except at its two ends, whenever it reaches a dot from one direction it leaves by another. Therefore the total number of edges meeting each point is *even* – except for the two ends, where it might just be odd.

However, for the bridges, these numbers are 3, 3, 3, and 5: *all* odd. Therefore no tour is possible.

This gives a necessary condition for a complete tour. At most two dots must lie on an odd number of edges. Euler proved that it's also sufficient. If at most two dots lie on an odd number of edges, then a tour exists. It must start and end at the dots with an odd number of edges, if there are any. If there aren't, it can start anywhere; moreover, it can then be closed up into a loop that starts and ends at the same place. The proof isn't especially hard, but it takes a little setting up, so I won't give it.

Some years after Euler's work, traffic in Königsberg got so heavy that the city fathers built the Kneiphof relief road (figure 4.3(b)). The numbers of lines meeting each dot were then 6, 3, 3, 4. Exactly two are odd; so by Euler's theorem it's possible to find a tour. The tour must start on the North bank and end on the South – or vice versa. Such a tour is shown. Soon, instead of walking the famous route, the good citizens of Königsberg began to drive round it after lunch every Sunday.

In consequence, the traffic grew worse, and in desperation the city fathers introduced the Königsberg one-way system (figure 4.3(c)). Several eminent citizens were fined for following the tour shown in figure 4.3(b), not having noticed that it went the wrong way up the final street. They then began to ask whether a legal tour was possible, and discovered that Euler had already thought of this too.

A one-way system corresponds to a *directed* graph in which each edge is marked with an arrow, and must be traversed in the direction of the arrow. Again, Euler asked what would happen if a tour existed. At each dot, other than the ends, the path must enter and then leave. Thus the number of arrows coming *into* the dot must equal the number going *out*. At the ends of the tour, there must be one dot having one more entrance than it has exits, and one having one more exit than it has entrances. These conditions are also sufficient for a tour; and if all dots have the same number of entrances as they do exits, a circular tour is possible.

It's easy to see that figure 4.3(c) violates Euler's conditions. In fact, once you've crossed the leftmost bridge going north, there are two southbound bridges. You can only use one of them, and you can never get back to use the other. This ruined the inhabitants' Sunday drives, and after a petition was delivered to City Hall, the system was changed to the one in figure 4.3(d), satisfying Euler's conditions.

Back to ouroborean rings.

Let's find one for quadruples. The Good idea is to represent each quadruple as a one-way street leading from its initial triple to its final triple. For instance 0110 is the road from town 011 to town 110, and is one way in that direction. There are eight triples, so there are eight towns joined by sixteen roads. The corresponding graph is shown in figure 4.4.

It satisfies Euler's conditions. At each town, two roads enter and two exit. Therefore there exists a circular tour; but that is an ouroborean ring.

The same argument works for quintuples, sextuples, and so on. And you can *see* why Euler's conditions have to be satisfied. For example, the roads leaving town 001 are either 0010 or 0011 – the town's name, plus either 0 or 1. That's two exits. By the same argument, there are two entrances: 0001 and 1001.

Longer ouroborean rings, and their mathematical relatives, are used by electronic engineers to code messages. The 0's and 1's are binary digits: 1 is a pulse of electricity, 0 the absence of a pulse. Such codes have applications to telephone transmission and radar-mapping. The surface of Venus has been mapped from Earth by radar! Paradoxically, the returning signal is so weak that on average less than one quantum of energy returns. But the quantum is the smallest possible unit of energy! The answer to this paradox is that the coding method is very highly redundant: most digits can be missing and the signal still makes sense. So when the odd quantum gets lucky and returns to Earth, it contributes to a meaningful signal.

It takes ages.

Stein's book includes a tabulation of the history of ouroborean rings, from AD 1000 until 1960. The book deals only with sequences of 0's and 1's; but we can ask the same questions for, say, pairs of digits 0, 1, 2. There are nine of these. The Good Road Guide to the nine towns of the province of Pairs-from-Three looks like figure 4.5. Three roads run into each town, and the same number leave, so Euler's theorem tells us that a circular tour is possible. The one shown gives the ouroborean ring

001122102.

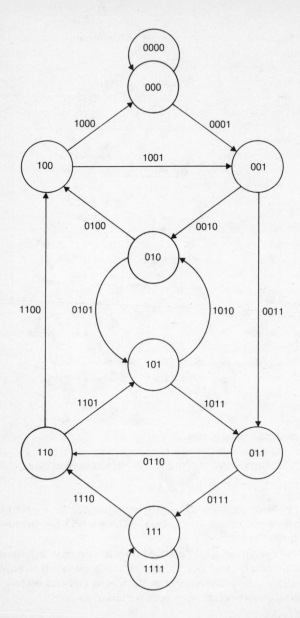

4.4 *The Good Road Guide to triples.*

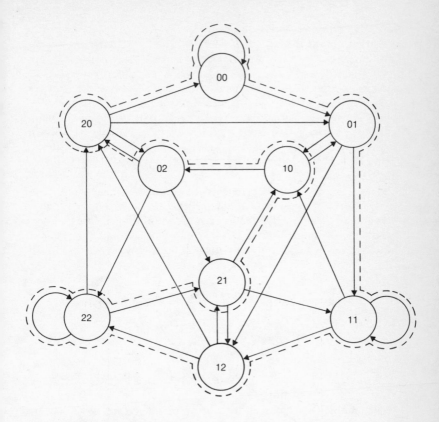

4.5   *The Good Road Guide to three-digit pairs.*

Similarly, for three-digit triples there is an ouroborean ring

000111222121102202101201002.

In fact I got these sequences by applying an algorithm – a step-by-step method whose success is guaranteed – which works for any *m*-digit *n*-tuples. Here's the method.

Consider the case of triples formed from the digits 0, 1, 2: the general case is entirely similar. Start by writing out a list of all twenty-seven possible triples, in numerical order, starting at 000 and ending at 222. Now write down the start of an ouroborean ring:

000111222

and delete from your list all triples included in it (so, for example, you delete 000, 001, 011, 111, and so on up to 222). Now look at the *largest* triple beginning 22: it is 221. Write down the 1 on the end of the ouroborean ring and cross 221 off the list. Now look for the largest triple starting 21. Repeat, *always using the largest available triple* and then crossing it off. You don't get stuck, and it closes up. The result is an ouroborean ring.

As I said, the algorithm is guaranteed to work. A proof was given by M. H. Martin in 1934.

Armed with this method, you will now be able to write down, before breakfast and with your hands tied behind your back, an ouroborean ring of length 117,649 containing every possible seven-digit sextuplet. Or, if you're not so ambitious, every four-digit pair or every three-digit quadruplet. *Go on, try it!*

Martin's algorithm only produces one ouroborean ring for any given m and n, but in general there are many other solutions. For two-digit n-tuplets there is a formula for the number of ouroborean rings, found by N. G. de Bruijn in 1946: it is *two raised to the power* $2^{n-1} - n$, which grows extremely fast. Here rings obtained by revolving a given one are considered the same. Here's a table:

n	Number of ouroborean rings
2	1
3	2
4	16
5	2048
6	67108864
7	144115188075855872

The mathematical possibilities of autovoracity are by no means exhausted. Are there higher-dimensional analogues of ouroborean rings? For example, there are sixteen 2×2 squares with entries 0 or 1. Is it possible to write 0's and 1's in a 4×4 square so that the sixteen subsquares list each possibility exactly once? You must pretend that opposite edges of the square are joined together, so that it wraps round into a torus: I call this the *Ourotorus Problem*.

Here's a different way to say it. Cut out the sixteen pieces shown in figure 4.6. Note the white dot near the top to tell you which way up they go. Can you arrange them in a 4×4 grid, keeping the dot at the top, so that adjacent squares have the same shade along common edges? Again, this rule also applies to squares that become adjacent if the top and bottom of the grid are joined, or if the left and right sides are.

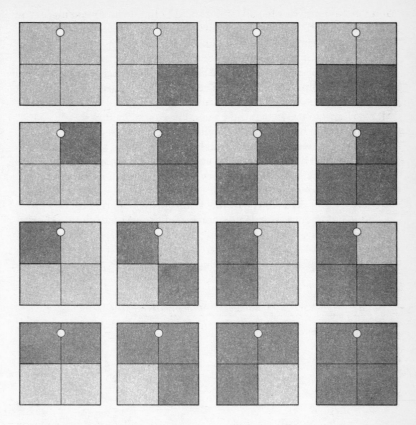

4.6 Pieces for the Ourotorus Problem.

I don't want to deprive you of the pleasure of looking for the solution, so I'll tell you what it is at the end of this chapter. I'll show you something else instead.

There are eighty-one possible 2×2 squares containing the digits 0, 1, 2. I can fit them together to make a 9×9 ourotorus, as follows. Begin with the corresponding ouroborean ring

001122102.

Write this down again:

001122102.

Then shift it one space to the right, remembering to wrap the last digit round to the start:

200112210.

Next shift this two spaces right; shift the result three spaces right; and so on, each shift being one space more than the previous one. The result, shown in figure 4.7, is an ourotorus!

It's not hard to prove that this method works, *without* checking every subsquare. (*Hint*: think about the top and bottom rows of the 2×2 subsquare.) The same method of construction works for m-digit 2×2 squares *whenever* m *is odd*. So, for example, you can generate a five-digit ourotorus of 2×2 squares by repeatedly shifting an ouroborean ring made up of five-digit pairs.

0	0	1	1	2	2	1	0	2
0	0	1	1	2	2	1	0	2
2	0	0	1	1	2	2	1	0
1	0	2	0	0	1	1	2	2
1	2	2	1	0	2	0	0	1
2	0	0	1	1	2	2	1	0
1	2	2	1	0	2	0	0	1
1	0	2	0	0	1	1	2	2
2	0	0	1	1	2	2	1	0

4.7 *An ourotorus for three-digit pairs.*

But the method fails when m is even. In fact, I don't know whether there exists an ourotorus for four-digit 2×2 squares. Can anyone find one? Are there any general methods for obtaining ourotori for m-digit 2×2 squares when m is even?

Every row of an ourotorus constructed by this method is an ouroborean ring. Does there exist an ourotorus for three-digit 2×2 squares all of whose *columns* are also ouroborean rings? My solution doesn't have this property.

I also have no results for 3×3 squares, say, except for some obvious remarks. The number of two-digit 3×3 squares is 2^9, which is not a square. So a "square" ourotorus can't exist in this case. However, there might be a rectangular one, say 16×32. Similar remarks apply to $m \times m$ squares with m odd, unless the number of digits is itself a square.

And what about three dimensions? There are $2^8 = 256$ $2 \times 2 \times 2$ cubes containing 0's and 1's. Can these be obtained from all the subcubes of a cube? No, because 256 isn't a cube. However, there are $2^{27} = 134{,}217{,}728$ two-digit $3 \times 3 \times 3$ cubes, and that's the cube of 512 . . .

The mind boggles. *Yamátárájabhánasalagám* . . .

ANSWERS

Four-digit pairs: using Martin's algorithm, you get

>   0011223321310302

Three-digit quadruples: the same method yields

>   00001111222212211212111022202210212021101220121011120110022
>   00210012001020201010002

A 4×4 ourotorus (it is essentially unique) is shown in figure 4. 8. If its design is repeated as shown, you get a remarkable tiling of the plane by cross-shaped dark and light tiles, in which all possible 2×2 arrays of dark and light squares appear in a regular manner.

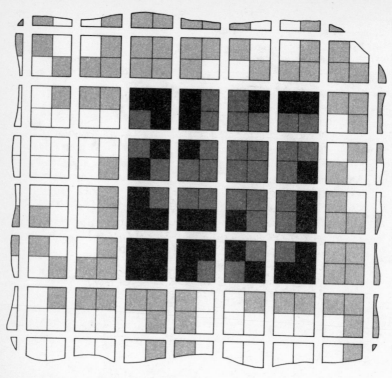

4.8 *A 4×4 ourotorus and its associated tiling.*

FURTHER READING

N. G. de Bruijn, "A Combinatorial Problem", *Akademie van Wetenschappen* (Amsterdam), 8 (1946), pp. 461–7

I. J. Good, "Normal Recurring Decimals", *Journal of the London Mathematical Society*, 21 (1946), pp. 167–9

M. H. Martin, "A Problem in Arrangements", *Bulletin of the American Mathematical Society*, 40 (1934), pp. 859–64

G. H. Pettengill, D. B. Campbell, and H. Masursky, "The Surface of Venus", *Scientific American* (August 1980), pp. 46–57

Manfred Schroeder, *Number Theory in Science and Communication* (New York: Springer, 1984)

Sherman K. Stein, "The Mathematician as an Explorer", *Scientific American* (May 1961), pp. 149–58

Sherman K. Stein, *Mathematics: the Man-Made Universe* (San Francisco: Freeman, 1976)

5

Fallacy or Ycallaf?

It was a dark, cold, winter evening in Logicland. Tweedledumb and Tweedledim, the Terrible Twins, were having a logical debate. In other words, they were arguing.

As usual.

"Only an elephant or a whale gives birth to a creature that weighs more than 100 kilograms," said Tweedledumb. "Right?"

"I suppose," said Tweedledim.

"The President weighs 101 kilograms," said Tweedledumb.

"Uh-oh," said Tweedledim. "I can see exactly where you're heading and I don't . . ."

"Therefore," said Tweedledumb . . .

"The President's mother is either an elephant or a whale!" they yelled in chorus.

"That's a fallacy!" screamed Tweedledim.

"What's a fallacy?"

"A *fallacy* is an apparently convincing argument that is logically false," said Tweedledim. "I'm surprised you didn't know . . ."

"Of course I know what a fallacy is! I meant, which step in my deduction is fallacious?"

"The first. No, the second. No, they're both correct, but you've forgotten that . . ."

"See? It's not a fallacy at all! It's an *ycallaf* ! "

"What's an ycallaf?"

"My logical deduction about the . . ."

"No, 'Dumb! I know you're referring to your deduction. I mean, what on earth is an 'ycallaf'?"

"An *ycallaf*," said Tweedledumb, "is an apparently false argument that is actually logically correct."

"Your argument isn't an ycallaf! It's a fallacy!"

"Isn't!"

"Is!"

"Isn't!"

"Is!"

"Isn't!"

"Is! Is! Is! Is!"

They carried on like that for some time. There's nothing like a good logical argument – and what they had *was* nothing like a good logical argument.

Everyone in Logicland is either a logician or a mathematician. It's a funny place. You see, to do mathematics you have to be good at logic. In fact mathematical research is the art of telling the fallacies from the ycallafs. How good would you be at research mathematics? Here are ten problems to test your aptitude. All you have to do is decide which is which. Answers at the end. Good luck!

1 Tangram Twins

The old Chinese puzzle of tangrams consists of a square, cut into seven pieces. Tweedledumb and Tweedledim were playing tangrams.

"I've made a man," said Tweedledumb.

"So have I," said Tweedledim (figure 5.1).

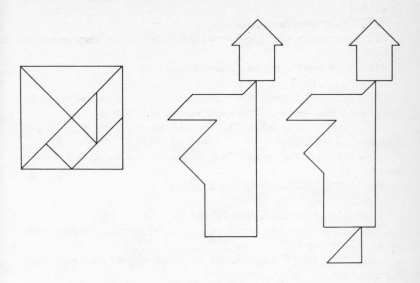

5.1 *Tangrams – has Tweedledim missed a piece out?*

"You idiot, yours hasn't got a foot!" said Tweedledumb. "You must have missed out one of the pieces."

Fallacy or ycallaf?

2 Logarithmania

"You like mathematics, don't you, 'Dumb?" said Tweedledim.

"Provided it's accompanied by a good wine and served hot . . ."

"You'll love this, then! Now, you'll remember that the logarithmic series

$$\log (1 + x) = x - \tfrac{1}{2}x^2 + \tfrac{1}{3}x^3 - \dots$$

is valid provided x is greater than -1 and less than or equal to 1."

"Convergent, you mean."

"Precisely! You are in an agreeable mood today, 'Dumb!"

"No I'm not!"

"Yes you are!"

"No I'm . . ."

"Have it your own way. Now, put $x = 1$ to get

$$\log 2 = 1 - \tfrac{1}{2} + \tfrac{1}{3} - \tfrac{1}{4} + \tfrac{1}{5} - \tfrac{1}{6} + \tfrac{1}{7} - \tfrac{1}{8} + \tfrac{1}{9} - \dots.$$

Double it:

$$2 \log 2 = 2 - \tfrac{2}{2} + \tfrac{2}{3} - \tfrac{2}{4} + \tfrac{2}{5} - \tfrac{2}{6} + \tfrac{2}{7} - \tfrac{2}{8} + \tfrac{2}{9} - \ldots$$

$$= 2 - 1 + \tfrac{2}{3} - \tfrac{1}{2} + \tfrac{2}{5} - \tfrac{1}{3} + \tfrac{2}{7} - \tfrac{1}{4} + \tfrac{2}{9} - \ldots$$

Collect pairs of terms with the same denominator. Now you get

$$2 \log 2 = 1 - \tfrac{1}{2} + \tfrac{1}{3} - \tfrac{1}{4} + \tfrac{1}{5} - \tfrac{1}{6} + \tfrac{1}{7} - \tfrac{1}{8} + \tfrac{1}{9} - \ldots$$

$$= \log 2.$$

Therefore $2 \log 2 = \log 2$, that is, $2 = 1$. Neat, isn't it?"
   Fallacy or ycallaf?

3   Easy for Sum

"I know one like that," said Tweedledumb. "Take the series

$$1 - 1 + 1 - 1 + 1 - 1 + 1 - 1 + \ldots.$$

Bracketed like this,

$$(1 - 1) + (1 - 1) + (1 - 1) + \ldots$$

the sum is 0. But bracketed as

$$1 + (-1 + 1) + (-1 + 1) + (-1 + 1) + \ldots$$

the sum is 1. So $1 = 0$. Incidentally, that confirms *your* result: just add 1 to each side! "
   Fallacy or ycallaf?

4   Knot so Easy

Tweedledim fancies himself as a conjuror. "Hey, 'Dumb! Here's a good trick! First, I tie a knot in this string, like so . . . Then I tie another one . . . Abracasesame! Look, they've both vanished!" (Figure 5.2)

   "That's silly, 'Dim. All you've done is tie a knot and its antiknot, so that they both cancel out."

   "*Antiknot?* Who ever heard of an antiknot?"

   "The same knot, tied inside out."

   "What rubbish! There's no such thing as an antiknot! And I can prove it! Did you know that you can do arithmetic with knots? To add two knots K and L you just tie them in turn in the same piece of string (figure 5.3). Call the result $K+L$, right?"

   "If you insist."

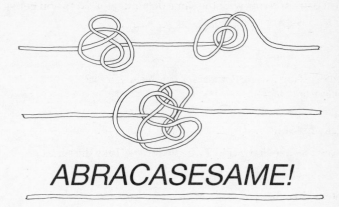

ABRACASESAME!

5.2 *Two knots . . . Merge them . . . They cancel. Can it be done?*

"Good. Now, obviously 0 must be the unknot – a knot that isn't knotted, if you see what I mean."

"Why?"

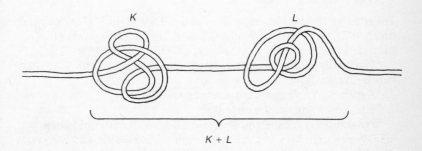

5.3 *Knot arithmetic.*

"Well, if you tie a knot K and then tie the unknot, it's the same as just tying K, so $K + 0 = K$. It makes sense. Now, if K is a knot, then its antiknot can be written as $-K$, because $K + (-K)$ has to equal 0."

"Aha! So you agree that an antiknot can exist!"

"No, no. I'm going to show that the only knot that has an antiknot is the unknot."

"Oh. What? Say that ag- . . ."

5.4 *An infinite knot that proves cancellation is impossible. Or does it?*

"Wash your ears, you'll hear better! Now, suppose I tie the infinite knot (figure 5.4)

$$K - K + K - K + K - K + \ldots.$$"

"You know, I've seen something like this bef- . . ."

"Bracketed like this,

$$(K - K) + (K - K) + (K - K) + \ldots$$

its value is 0. But, bracketed like *this*:

$$K + (-K + K) + (-K + K) + (-K + K) + \ldots,$$

its value is . . ."

"*K*. So *K* = 0. Yeah, sure. And you accused *me* of talking rubbish!"

Fallacy or ycallaf?

5 Poker by Phone

Tweedledumb and Tweedledim enjoy a good game of cards. But Tweedledim is going on holiday soon.

"I'll miss our card game."

"Me too. We have such vicious fights . . ."

"I know! We'll play poker by phone! I'll deal out the cards, send you your five, and then we'll tell each other which card we're playing."

Tweedledumb considered this. "Great idea," he said, with heavy sarcasm. "How do I know you won't cheat?"

"I promise not to."

"Liar! You *always* tell lies!"

"Yes, I do. Anyway, how do I know *you* won't cheat?"

"We can tell each other what all the cards are," said Tweedledumb. They thought about this for a moment.

"That's stupid!"

"No it's not. We can put them in code, so that the other one can't decode them. Then, at the end, we can reveal our codes and check that nobody's cheated!"

"I'm sorry, 'Dumb, but I really don't see what you're on about."

"OK. Listen very carefully, I shall say this only once . . . You've heard of trapdoor codes?"

"Theoretically unbreakable codes? Where you can tell anyone how to put a message *into* code, but that doesn't help them *decode* anything?"

"You got it. Now, you choose an encoding rule E_{dim} and a decoding rule D_{dim}; I choose rules E_{dumb} and D_{dumb}. We both know the Es, but we only know our own Ds."

"OK so far."

"If I put a message M into code then I get $E_{dumb}M$. To decode it, I work out $D_{dumb}E_{dumb}M$, which is M. So D_{dumb} undoes E_{dumb}. Now, I take the fifty-two messages . . ."

"*I* take the fifty-two messages!"

"Very well. *You* take the fifty-two messages

ACE OF CLUBS

TWO OF CLUBS . . .

and so on up to

KING OF SPADES.

You put each one into code using your rule. Each message M is changed to $E_{dim}M$. You shuffle the lot . . ."

"Randomly rearrange them, you mean?"

"Right. Then you transmit the lot to me."

"Sounds fair enough so far."

"I then select five messages at random, and send them back to you so that you can decode them to find out what your hand is. I can't know your hand, because I can't decode your encoding rule. Then, I select five more random messages, constituting my hand."

"Ah, but you've got a problem then. You can't decode them to find what they are!"

"No, but I'm clever. I encode them a second time, using my rule E_{dumb}. So if the message is M, it becomes $E_{dumb}E_{dim}M$. Then I send them back to *you*, and you undo the effect of your code E_{dim} by applying D_{dim}. That gives $D_{dim}E_{dumb}E_{dim}M$, which is the same as $E_{dumb}M$."

"You're tacitly assuming that $E_{dumb}E_{dim} = E_{dim}E_{dumb}$."

"Oh, so I am. But that can be arranged if we choose the right codes. Let's suppose it has. You send the five messages $E_{dumb}M$ back to me. I decode them by applying D_{dumb}. Now we've each got a hand of cards, with no cards common to both hands, and neither knows what cards the other has, so we can play. We keep a tape of all messages, and at the end, we each reveal our decoding rules, so that we can both check nobody's cheated at any stage."

Tweedledim considered this at length. "Heck, it's complicated," he said.

"Look, here's an analogy. Poker by post. You put the fifty-two cards into identical boxes, and padlock them all with locks *to which only you have the key*. You send them to me. I select five at random to make up your hand. I select another five at random for my hand, and put another padlock on those, to which only I have the key. I send all ten to you; you take off your padlocks and send my five back to me, with my locks still in place. Simple!"

Tweedledim got out a pad and started scribbling on it, checking the logic. Suddenly he stopped.

"Hang on," he said. "Suppose there are just *three* cards."

"But there are fifty-two."

"Yes, but the method ought to work with just three. Now, we send each other lots of messages, at the end of which we each know one card – our own – *and* we know they're different. Right?"

"Right."

"OK. Now, let S_{dim} be the set of cards that I could have ended up with, consistent with those messages, and let S_{dumb} be the set of cards you might have ended up with. Then my card belongs to S_{dim}, and yours belongs to S_{dumb}."

"I begin to see your drift," said Tweedledumb. "Either of us can work out S_{dim} and S_{dumb} by pure logic. That's the whole point. So the set S_{dim} can't be *just* your card."

"No. Otherwise you'd know what my card was. On the other hand, S_{dim} and S_{dumb} can't contain a common card, or else we might both be getting the same card. So S_{dim} can't be all three cards, because then you can't get any cards at all."

"I see. So S_{dim} contains exactly two cards out of the three."

"Excellent! But, by the same token, so does S_{dumb}."

"And the sets don't overlap, so there have to be at least four cards altogether," said Tweedledumb. "But, there are only three."

"So we can't play poker by telephone after all," said Tweedledim.
Fallacy or ycallaf?

6 Was Galileo Right?

During a particularly violent argument, about the smallest whole number that cannot be described using less than fourteen words, Tweedledumb picked up a teacup and threw it at Tweedledim.

"Yah! Missed! Can't you even compute a parabolic arc?"

"What's a parabola got to do with it?"

"Galileo proved that the path of a falling projectile is a parabola."

"No it's not."

"Neglecting air resistance, of course."

"It still isn't. Galileo got it wrong."
Fallacy or Ycallaf?

7 Au Courant

Tweedledim had been reading a classic of expository mathematics, *What is Mathematics?* by Richard Courant and Herbert Robbins.

"Hey, here's a good one! Wake up, 'Dumb!"

"Gronfff. What?"

"Suppose a train travels between two railway stations along a straight track. A rod is hinged to the floor of one of the carriages, able to move without friction either forward or backward until it touches the floor (figure 5.5). If it does touch the floor, assume it stays there throughout the subsequent motion. Suppose I specify *in advance* how the train moves. But the motion need not be uniform: the train can speed up, stop suddenly, even go into reverse for a time. It must start at one station and end at the other. Can you always place the rod in such a position that it never hits the floor during the journey?"

"Hmmph . . . Tricky. The equations of motion are . . . Oh, wait, I get it! It's a topological problem!"

"Eh? What have rubber sheets got to do with . . ."

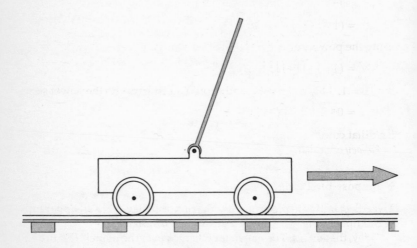

5.5 *Hinged rod on a railway carriage. Can it fail to hit the floor?*

"No, no. It's about *continuity*. The final position of the rod depends continuously on its initial position! Now, there's a continuous range of angles that I can start it at, from 0° to 180°, so the range of final angles is also continuous. If I start it lying down forwards at 0°, it stays there. If I start it lying down backwards at 180°, it stays *there*. So the range of final angles includes all values between 0° and 180°. In particular, it includes 90°, so I can arrange for the rod to finish up vertical. Since it stays on the floor when it hits it, it can't hit the floor at all."

Fallacy or ycallaf?

8 Integral Equation

"I've got another mathematical one," said Tweedledim.

"Show-off."

"It's a calculus question. You know that when you integrate the exponential function e^x you just get e^x again?"

"You mean, $\int e^x = e^x$?"

"Right! Now, write that as

$$(1 - \int)e^x = 0,$$

so

$$e^x = \frac{1}{1-\int}0$$

$$= (1 + \int + \int^2 + \int^3 + \ldots)0$$

using the power series for $\frac{1}{1-\int}$. In other words,

$$e^x = (1 + \int + \int\int + \int\int\int + \ldots)0.$$

But $\int 0 = 1$, $\int 1 = x$, $\int x = \frac{1}{2}x^2$, and so on. You end up with the power series

$$e^x = 0 + 1 + x + \frac{1}{2}x^2 + \frac{1}{6}x^3 + \ldots .$$

Isn't that cute?"

Fallacy or ycallaf?

9 Impossible Tiling?

Tweedledumb was playing with lots of tiles. They were all apparently regular polygons, with equal sides and equal angles.

"Hey, that's neat! They all fit together to cover the plane!" (Figure 5.6)

"Let me look at that," said Tweedledim. "Something's wrong, 'Dumb! If you tile the whole plane with a mixture of equal-sided regular polygons you can do it using polygons with three, four, six, eight, and twelve sides, but no others. But your tiling has got polygons with five and seven sides in it! You must have made a mistake."

"Well, look for yourself!"

Fallacy or ycallaf?

10 Spelling Mistakes

"My turn," said Tweedledim. "A quickie to end on.

"Ther are five mistokes im this centence."

Fallacy or ycallaf?

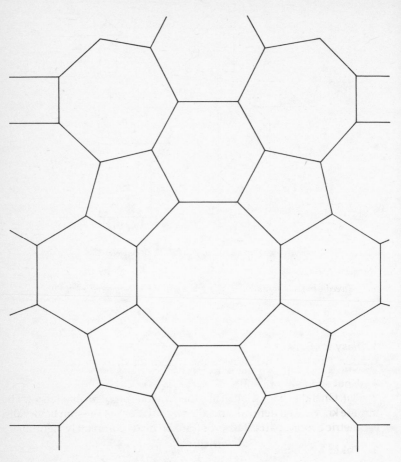

5.6   A new tiling by regular polygons?

ANSWERS

1   Tangram Twins.

Fallacy. Tweedledim hasn't missed out a piece. He's just found a different arrangement (figure 5.7).

2   Logarithmania

Fallacy. The logarithmic series is not *absolutely* convergent (convergent if every term is made positive) and hence cannot be rearranged.

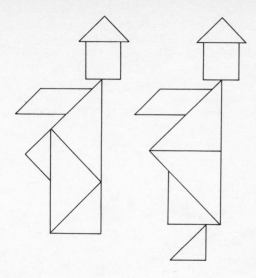

5.7   *Tweedledim's tangram man, and Tweedledumb's. No pieces missing!*

3   Easy for Sum
Fallacy. The sum of the series is not well defined.

4   Knot so Easy
Ycallaf. On the surface it looks like the previous problem, but it isn't. The infinite knot *is* well defined, and all manipulations of the sum have valid geometric counterparts. The proof can be made completely rigorous!

5   Poker by Phone
Fallacy *and* ycallaf! Both arguments are "right". The second (impossibility proof) does not contradict the first (practical solution). The point is that *given a sufficiently long time* the code messages involved can be decoded, and the poker game then becomes impossible. But in practice the time would be longer than the age of the universe. That's an "unbreakable code" for practical purposes. For more about this kind of code, see the items in "Further Reading" by Gardner, Hellman, and Klarner.

6   Was Galileo Right?
Ycallaf. Galileo's result assumes a flat Earth and constant gravity. With a spherical Earth, and Newtonian gravity, the path of a falling body is like that of any other body revolving round the Earth: an ellipse.

7   Au Courant

Fallacy. (I'm going to get lots of letters of protest!) *If* the assumption of continuity is correct, the argument is an ycallaf. *But* the continuity assumption is *not* justified. The problem is those "absorbing boundary conditions": *if the rod hits the floor, then it stays there.*

Imagine first that the rod can turn a full 360° – no floor. Then a possible history is shown in figure 5.8(a). When the absorbing boundary conditions are put back (figure 5.8(b)), all initial positions end up on the floor.

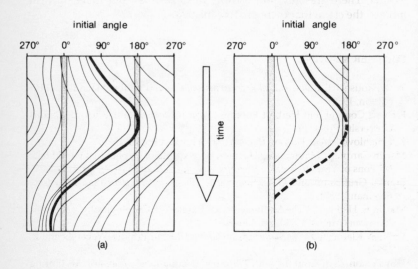

5.8   Why Courant and Robbins should not have assumed continuity. The graphs show the histories of various initial positions. The two pictures are identical, except that in (b) the "absorbing boundary conditions" have been added (grey lines). The history shown as a heavy line causes no trouble in (a). However, because it is tangent to the grey line at 180°, representing the floor, in (b) it causes a discontinuity. Any initial position to the left of the heavy line ends up on the floor at 0°; any initial position to the right of, or on, the heavy line ends up at 180°.

This error in Courant and Robbins's reasoning was first pointed out by Tim Poston in *Manifold* magazine in 1976. It's still not as widely known as it should be.

8   Integral Equation

Ycallaf. The theory of Banach spaces can be used to give a rigorous justification for the manipulations of the $\int$ sign. If it is treated as an operator, the series expansion for $\frac{1}{(1-\int)}$ is correct!

9 Impossible Tiling?

Fallacy. The tiles are either not exactly regular polygons, or they don't meet exactly. For example, one vertex is surrounded by a pentagon, a hexagon, and an octagon. If these are regular, their angles are 108°, 120°, and 135°, which add up to 363°. But if the tiles fit exactly, the sum should be 360°.

The pattern is taken from an ancient Islamic design.

10 Spelling Mistakes

Ycallaf. There are only four *spelling* mistakes, yes. But there's a fifth *mistake*: the claim that there are five mistakes!

FURTHER READING

W. W. Rouse Ball, *Mathematical Recreations and Essays* (London: Macmillan, 11th edition, 1959)

Richard Courant and Herbert Robbins, *What is Mathematics?* (Oxford: Oxford University Press, 1941)

K. Critchlow, *Islamic Patterns* (New York: Schocken Books, 1976)

Martin Gardner, "Mathematical Games: a New Kind of Cypher that would take Millions of Years to Break", *Scientific American* (August 1977), pp. 120–4

Branko Grünbaum and G. C. Shephard, *Tilings and Patterns* (San Francisco: Freeman, 1987)

Martin E. Hellman, "The Mathematics of Public-key Cryptography", *Scientific American* (August 1979), pp. 130–9

David A. Klarner (ed.) *The Mathematical Gardner* (Boston: Prindle–Weber–Schmidt, 1981)

Tim Poston, "Au Courant with Differential Equations", *Manifold*, 18 (Spring 1976), pp. 6–9

6

Build your own Virus

Sniff.

I sat miserably in the doctor's waiting-room. Next to me was a rather large woman with a rather small child wrapped tightly in a crocheted shawl. The tiny pink face was covered in even tinier pink spots. I moved two seats to my left and tried to remember whether I'd already had chickenpox.

"Next!"

I caught the receptionist's eye and passed through to the inner sanctum.

I confess that I seldom really enjoy a visit to Dr Athanasius Fell; but despite his rather crusty exterior he is one of the best doctors around. Unfortunately he has a hatred of mathematics – a failing not unheard of among the medical profession – and he knows that I'm a mathematician. Our relationship is a little nervous.

"Hmmmph," he said. "You again."

"I wouldn't have bothered you but I've got a touch of influenza and I . . ."

"Mathematician." He made it sound like Typhoid Mary. "You can't fool me, I remember you. Parasites."

"No, it's influ- . . ."

"I don't mean you've *got* parasites. I mean mathematicians *are* parasites. No offence intended, you understand. Nothing personal. Just can't abide the creatures. Mathematicians, that is. What has mathematics ever done for medicine?"

What has medicine, I thought, *ever done for mathematics?* I was about to catalogue a range of medical applications of my subject, from the statistical analysis of epidemics to irregularities of the human heartbeat and the malfunctioning of the thyroid gland, but he cut me short. "Nothing," he said, in answer to his own rhetorical question.

Some devil within me – a phrase best associated with the Middle Ages rather than the modern practitioner, but one that remains curiously apt – prompted me to defend my profession. "You'd be surprised," I said.

He stuck a wooden stick in my mouth and inspected my tongue. "Hmmph. What?"

"It'th a funny thubject, mathematicth," I said. He gave me a look which suggested that this information was not novel, and removed the stick. I continued with greater clarity of diction. "Even the purest of pure mathematics has many unexpected applications."

"I've heard that excuse before."

"But it's true. You know about the ancient Greeks, of course . . ."

"Hippocrates was a Greek," he pointed out. "You may assume that I am conversant with the historical period."

"Yes, well . . . Now Euclid, you see . . ."

"Geometry." He made it sound like terminal gout.

"Right!" I said brightly. "The climax to the entire ten-volume work of Euclid's *Elements* is the proof that there are exactly five regular solids. The tetrahedron, cube, octahedron, dodecahedron, and . . ."

"Icosahedron," he finished for me. "You've got exophthalmia – protruding eyes," he observed. Ordinarily that's a disease of goldfish, but he was right. My eyes were protruding because he'd known about the icosahedron. "Regular solid with twenty equilateral triangles for faces.

Don't look so surprised. I do know *some* mathematics." He snorted. "Just don't see any *point* to it, that's all."

"Funny thing, the icosahedron," I said. "It was discovered somewhere around 370 BC, as a purely mathematical construct. At the time, nobody could find it in nature."

"Crystals," he said.

"Oddly enough, no," I replied. "Cubes, octahedra, and tetrahedra, yes. But you can't get fivefold symmetry in a crystal." I wondered whether I should tell him about the recent discovery of quasicrystals, which have a kind of short-range fivefold symmetry, but decided it would just confuse things.

"Footballs," he said.

I agreed that the modern soccer ball (figure 6.1) is essentially icosahedral in shape – in fact it's a *truncated* icosahedron, one whose corners have been cut off. I went on to explain that this had been selected because it was an excellent approximation to a sphere that could be made from plane patches of leather, and had replaced the older design based on a cube whose square faces were cut into three parallel rectangular strips. While he stuck a very cold stethoscope up my shirt I finally pointed out that the soccer ball had not existed in 370 BC.

"Pigs' bladders," he said. It was a colourful oath, one I hadn't heard before.

"No, I don't think . . ."

"Ah!" he said. "Biology course, years ago at Addenbrooke's . . . Radiolaria! Chap called Heckle, something like that."

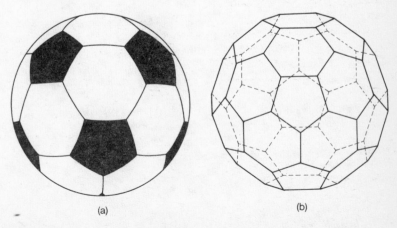

(a)   (b)

6.1 *The modern soccer ball (a) has the shape of a truncated icosahedron (b).*

Ernst Haeckel had made a long sea-voyage searching for scientific specimens, publishing his results in his *Challenger Monograph* of 1887. I knew this because it's mentioned in one of my favourite oddball books. "Yes, D'Arcy Thompson reproduced some of Haeckel's drawings in *On Growth and Form*." A radiolarian is a microscopic single-celled creature with a highly symmetric exterior skeleton. Haeckel had sketched hundreds of the things, and some were approximately icosahedral (figure 6.2). But – to be honest – there are reasons to suppose that Haeckel may have exaggerated the symmetry of his radiolaria a teensy bit. I pointed this out.

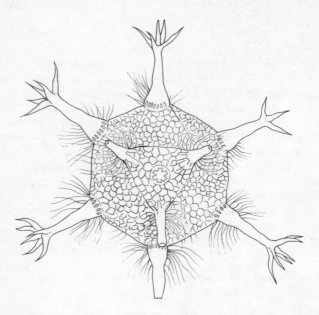

6.2 *The radiolarian* circogonia icosahedra *from Ernst Haeckel's* Challenger Monograph.

"I'll give you a hint," I said. "Some scientists have called the icosahedron 'Nature's favourite shape'."

"Hmmph." Dr Athanasius Fell scratched his beard. "I give up."

"Shame on you. And you a medical man, too."

"What's that supposed to mean?"

"Smallpox," I said.

"Eh?"

"Polio. Herpes. Turnip yellow mosaic ..."

He palpated my stomach, firmly enough to make me wince.

"The icosahedron," I continued doggedly, "is one of the commonest shapes for a virus."

That woke him up. "Is it? Let me see . . ." He dug out a huge textbook and thumbed its pages. "Good God, the mathematician's got it right for once! (figure 6.3). Now, why would anyone pick an icosahedron?"

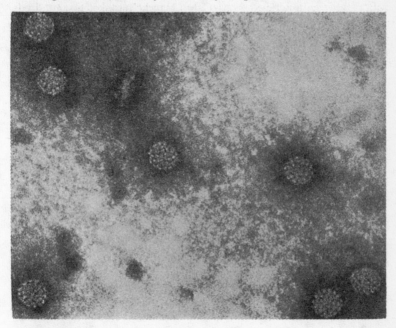

6.3   *The human wart virus is made up from seventy-two identical units arranged with icosahedral symmetry.* [Madeley, Virus Morphology]

"It may be the same reason that applies to soccer balls," I said. "If you want to make a roughly spherical body from a smallish number of identical units, then the icosahedron is the best shape. If you want a deeper explanation, it's probably that configurations with minimal energy tend to be symmetric, and . . ."

"No, no, that will do fine," he said quickly. "Mind you," he went on, thumbing his book again, "there are other virus shapes too."

In fact, the other most common shape for a virus is that of a helix (figure 6.4), a spiral wound to form a tube, like the thread on a screw or a spiral staircase. Again, this is presumably a minimum-energy configuration for identical units.

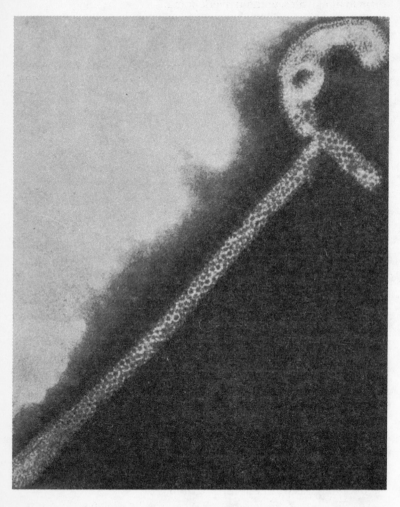

6.4　*The influenza C virus is a helix built from identical units arranged like treads in a spiral staircase.* [Madeley, Virus Morphology]

"Oh, look," he said. "Here's a helix made up of hexagonal components. Influenza C, the Taylor virus. Now, why doesn't that form something spherical instead of cylindrical?"

"You can't cover a sphere with hexagons," I said. "Not if you want them to fit like tiles, without overlapping."

"Why not?"

"Euler's theorem."

"Theorem?! Theorem?! Don't hold with those damned things at all, I'll tell you. Angles at the base of an isosceles triangle – hmmph, angles on the head of a pin! Pythagoras, silly ass. Impossible? Pah, nothing's impossible if you try hard enough! Euler? Sounds defeatist to me. Never heard of him anyway. "

"He's the most prolific mathematician of all time."

"That explains it."

"Euler proved that if F is the number of faces of a solid, V the number of vertices, and E the number of edges, then $F + V - E = 2$. Now, if you have a solid with F faces all hexagonal, then the number of edges must be

$$E = 3F$$

because each hexagon has six edges but every edge adjoins two hexagons; and the number of vertices is

$$V = 2F$$

because each face has six vertices but every vertex appears on exactly three neighbouring faces. So by Euler's theorem we have

$$F + 2F - 3F = 2$$

But in fact $F + 2F - 3F = 0$. So it can't be done."

For a moment he looked impressed, but he shook his head like a penguin shaking its feathers after a dip in the Southern Ocean, and the usual crusty visage reasserted itself. "Poliovirus, you said."

"Yes. A team of scientists at the Scripps clinic used X-ray crystallography to show that the poliovirus has much the same structure as a soccer ball (box 6.1). In fact, the general structure was suggested in 1962 by D. Caspar (Children's Cancer Research Foundation, Boston) and Aaron Klug (Laboratory of Molecular Biology, Cambridge) on mathematical grounds, but this wasn't confirmed until 1987."

Dr Fell, meanwhile, had returned to an earlier remark. "If you can't cover a sphere with hexagons, what about pentagons?"

Well, of course that's very interesting. Euler's formula again can be used to show that if you use only pentagons, then there have to be twelve of them, and the dodecahedron is the only possible shape. Now the problem is that the dodecahedron isn't really very rounded. Which is a pity, because spherical shapes tend to have the lowest energy – that's why a bubble or a raindrop is spherical. "Fortunately, you can get a

Box 6.1   How to make a model of the poliovirus

Copy figure 6.5(a) onto thin card, leaving a narrow border to make tabs on each edge. Turn it over and score along the boundaries between pentagons and around the outline.

Make twelve copies of figure 6.5(b) – a hexagon with one sector removed – also leaving spare borders on the edges for tabs. Turn over and score along the dotted lines. Fold and assemble into twelve five-sided pyramids, with flaps tucked under the base. Glue the flaps and stick the pyramids onto the white pentagons in figure 6.5(a). Then fold up the pentagons of figure 6.5(a) and glue to form a dodecahedron with pyramidal pimples on each face as in figure 6.5(c).

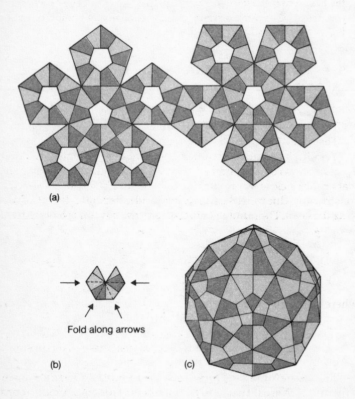

(a)

Fold along arrows

(b)   (c)

6.5   *Use (a) and twelve copies of (b) to make a model of the polio virus (c).*

rounder shape," I told him, "if you use only pentagons and hexagons. Then you have to have exactly *twelve* pentagons and the rest have to be hexagons . . ."

"Why? Why can't I take twenty-three pentagons plus enough hexagons?"

"Euler's formula again, though the argument is a little more involved (box 6.2)."

Box 6.2   Why there must be twelve pentagons

Suppose a polyhedron is made up from p pentagons and h hexagons, with no other faces. Then $F = p + h$. The pentagons have $5p$ edges, the hexagons $6h$, so the total number is $E = \frac{(5p + 6h)}{2}$ because each edge is counted twice, once for each face it adjoins. Similarly the number of vertices is $V = \frac{(5p + 6h)}{3}$. By Euler's formula,

$$2 = F + V - E = (p + h) + \frac{(5p + 6h)}{3} - \frac{(5p + 6h)}{2} = \frac{p}{6}.$$

So $p = 12$.

"Is there any limit on the number of hexagons?"

"The formula doesn't specify any."

"That's very curious."

"Yes, there's something rather strange about the pentagon. But there are restrictions on the number of hexagons if you want to make shapes that are very close approximations to a sphere." There's a very clever construction, due to Michael Goldberg and independently to Caspar and Klug (box 6.3). The number of faces, vertices, and edges have to be of the form

$F = 20T$ (twelve pentagons, the rest hexagons)
$E = 30T$
$V = 10T + 2$

where T is a number of the form

$T = a^2 + ab + b^2$.

This gives an "almost regular" solid, the pseudo-icosahedron of type $\{a,b\}$.

The "magic" numbers $10(a^2 + ab + b^2) + 2$ play a special role in the structure of a virus. These are the numbers of identical units (of protein molecules) that can be fitted together in an "almost regular" way to form a nearly spherical surface. Most numbers are not of this special form. As table 6.1 shows, the only magic numbers less than 300 are:

12, 32, 42, 72, 92, 122, 132, 162, 192, 212, 252, 272, 282.

Note that they all end in 2: can you see why? Their remainders on division by 3 appear to be either 0 or 2: is this correct, or can a number of the form $3k+1$ be magic? Why? Can a magic number be a perfect square?

Box 6.3 Goldberg–Caspar–Klug pseudo-icosahedra

Start with a tiling of the plane by equilateral triangles. Form large triangles as follows. Pick two numbers a and b. Starting at some vertex, move a units right and b units up to form a second vertex. Repeat to create a large triangle (figure 6.6(a)). Now fit together twenty of these large triangles, still divided up into the original smaller triangles, to form an icosahedron (figure 6.6(b)). This figure, projected from its centre onto a sphere, will yield a solid whose faces are very close to regular pentagons and hexagons, and having $V = 10\ (a^2 + ab + b^2) + 2$ vertices. This is the *pseudo-icosahedron* of type $\{a,b\}$.

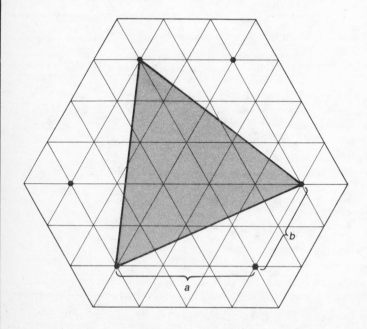

6.6 (a) Basic triangular unit of a pseudo-icosahedron of type $\{a,b\}$.

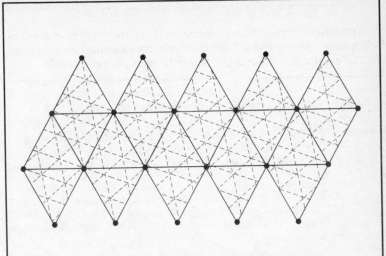

6.6 *(b) Assembling twenty basic units to make a pseudo-icosahedron.*

Table 6.1 Magic numbers for virus structure

a	b	$\left(a^2 + ab + b^2\right)$	$10\left(a^2 + ab + b^2\right) + 2$
1	0	1	12
1	1	3	32
2	0	4	42
2	1	7	72
2	2	12	122
3	0	9	92
3	1	13	132
3	2	19	192
3	3	27	272
4	0	16	162
4	1	21	212
4	2	28	282
5	0	25	252

"Totally mad," said Dr Fell. "You think that mathematics can dictate to Mother Nature? I'll tell you what you're suffering from, my boy. *Hubris*, that's what. And *Nemesis* will catch up with you, in the form of this book of viral data. Hmmmph . . . This should do it: *herpes simplex*, the cold sore virus – 162 units!"

"Type {4,0}," I said.

"What? Hmmph. So it is . . . Must be coincidence. Well, this next one will put a spoke in your wheel . . . Chicken adenovirus – 252 units."

"Type {5,0}." (Figure 6.7)

"Human wart!" he yelled. Several interpretations flashed across my mind. Another colourful insult? An oblique reference to the nose on his face? Not at all. He'd found another virus, and he wasn't any happier as a result. "Drat it! It's 72 – that's type {1,2}. BK virus – no, 72 as well. Rabbit papilloma – 72 again!"

"That's type {2,1}," I said. "Type {a,b} is the mirror image of type {b,a} when $a \neq b$. Oh, isn't that *amazing*! The human wart and the rabbit papilloma viruses are almost identical, except that one is left-handed and the other is right-handed! Don't you find that a fascinating example of convergent evolu- . . ."

"Turnip yellow mosaic – 32, type {1,1}! REO virus – bother, 92! Type {3,0}!" He flipped the pages like a man possessed. "Ah, a really big one, *bound* to go wrong . . . Infectious canine hepatitis! 362!" He looked at my list. "Not there, not there!"

"That," I said, "is because the list didn't go on far enough. Try type {6,0}."

He snorted uncomfortably.

"Even Nature," I said, "has to obey mathematical restrictions. *Provided* that the mathematics is an adequate description of Nature – which, of course, is often a moot question. But the combinatorics of repetitive structures is very basic, it's not really a surprise if it shows up in the real world.

"The same magic numbers turn up if you try to pack spheres together," I said. "If you start with one sphere you can pack twelve tightly round it. The next layer has forty-two spheres in it. Then there follow layers with 92, 162, 252, and so on. Buckminster Fuller thought that was very exciting. He suggested that 92 would have special mystical properties . . . For example, the ninety-second element is uranium. That's special, all right."

"Buckminster Fuller? Wasn't he an architect?"

"That's right. But he took a lot of inspiration from mathematics. He designed *geodesic domes* – spheres built up from triangular patterns (figure 6.8) – on the same principles as the Goldberg–Caspar–Klug virus shapes (table 6.2).

"The same ideas are important in chemistry too," I added, since the initiative was now firmly mine. "Chemists have synthesized organic molecules by joining up carbon atoms in the same kinds of pattern. They think the truncated icosahedron is especially important because it probably forms naturally in space, between the stars. They call the molecule

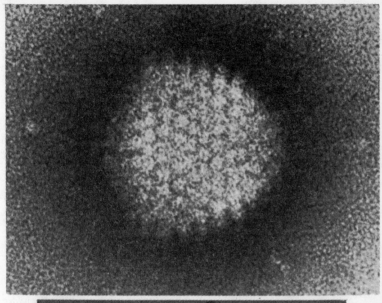

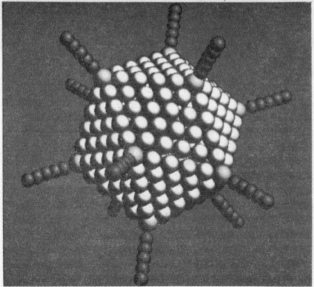

6.7   Adenovirus type 12, a type {5,0} arrangement of spherical protein units. To make a model like this all you need is 252 tennis balls (plus 60 for the spikes), several tubes of superglue, ingenuity, and persistence. [photographs, Science Photo Library]

6.8 *A geodesic dome. [photograph, Science Photo Library]*

Table 6.2 Pseudo-icosahedra: viruses and geodesic domes

{a,b}	virus	geodesic dome
{1,1}	turnip yellow mosaic	Arctic Institute, Baffin Island
{2,0}	bacteriophage ΦR	
{2,1}	rabbit papilloma	
{1,2}	human wart	
{2,2}		USS Leyte
{3,0}	REO	USAF Korea Officers' quarters
{4,0}	Herpes, chickenpox	Mount Washington
{5,0}	adenovirus type 12	US pavilion, Kabul
{6,0}	infectious canine hepatitis	Arctic DEW line radome
{8,8}		Lawrence, Long Island
{16,0}		US pavilion, Expo 67, Montreal
{18,0}		La Géode, Paris

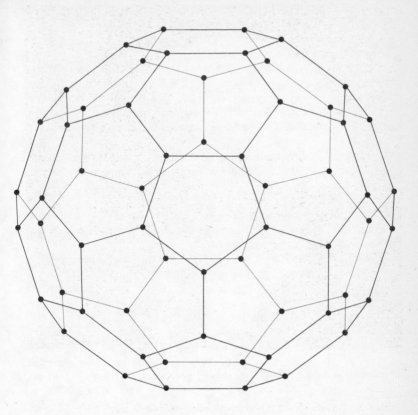

6.9 A cage of carbon atoms forms the basic skeleton of the buckminsterfullerene molecule. Compare with figure 6.1.

'Buckminsterfullerene' (figure 6.9) to honour the inventor of geodesic domes. Or sometimes 'footballene'. To – er – to honour the inventor of footballs . . ."

"*Buckminsterfullerene*," he said sadly. "*Footballene.*" His head drooped in defeat. But then I saw his eyes light up.

"I've diagnosed your complaint, young man," he said. "You're suffering from *pedodontia* – Foot-in-the-Mouth Disease – which, of course, is caused by a very *large* icosahedral virus! If not treated early, it leads to classic symptoms of a swollen head, which must be treated by installing a relief valve in the cranium. Now, the cure for a *large* virus is obviously a very *large* injection!" And he produced a syringe as long as my arm, looking like the sort of thing you spray trees with to kill swarms of locusts.

"I suddenly feel a lot better," I said.

"Nonsense! Just a quick jab . . ."

"Wait!" I yelled. "I've got a better idea!" I suddenly remembered a spoof conference report that had gone the rounds of mathematics departments a couple of years ago. It was a joke, of course, but maybe Dr Fell wouldn't realize that. "Wait! I've just remembered some work of Prof. Bertram Kostant at the Massachusetts Institute of Technology."

"So?"

"He used a mathematical analysis of the icosahedron to calculate its natural vibrational frequencies. What you need is a variable-frequency laser!"

"Why would I need a laser?"

"You could tune it to the precise frequency at which the virus will shake itself to bits! Like a wine-glass shattered by sympathetic vibrations if someone sings just the right note!"

"And what," he asked, "would I do with this laser if I had one?"

"Stick it up my nose and switch it on," I told him.

ANSWERS

Magic numbers are all of the form $10(a^2 + ab + b^2) + 2$, and since any multiple of 10 ends in the digit 0, magic numbers must end in the digit 2.

Magic numbers of the form $3k + 1$ cannot occur. To see why, calculate the possible values of $a^2 + ab + b^2$ (mod 3) :

a / b	0	1	2
0	0	1	1
1	1	0	1
2	1	1	0

We see that only the values 0 and 1 occur. Now any magic number $10(a^2 + ab + b^2) + 2$ becomes $1(a^2 + ab + b^2) + 2$ (mod 3), because $10 = 1$ (mod 3). This is just $a^2 + ab + b^2 + 2$, which is either $0 + 2 = 2$ or $1 + 2 = 0$ (mod 3).

No number ending in the digit 2 can be a perfect square, so square magic numbers don't exist.

FURTHER READING

H. S. M. Coxeter, "Virus Macromolecules and Geodesic Domes", *A Spectrum of Mathematics: Essays presented to H. G. Forder*, ed. John Butcher (Oxford: Oxford University Press, 1967)

H. M. Cundy and A. P. Rollett, *Mathematical Models* (Oxford: Clarendon Press, 1961)

James M. Hogle, Marie Chow, and David J. Filman, "The Structure of Poliovirus", *Scientific American* (March 1987), pp. 28–35

C. R. Madeley, *Virus Morphology* (Edinburgh: Churchill & Livingstone, 1972)

James Meller (ed.), *The Buckminster Fuller Reader* (Harmondsworth: Penguin Books, 1972)

Peter Pearce, *Structure in Nature is a Strategy for Design* (Boston: MIT Press, 1978)

D'Arcy W. Thompson, *On Growth and Form* (Cambridge: Cambridge University Press, 1942)

7

Parity Piece

I looked out of my window and a large tree flashed past it at more than a hundred kilometres per hour. But that's perfectly normal when you're on a train. The compartment was nearly empty. Apart from myself there was only a man in black trousers, black sandals, and bare feet. I couldn't see the rest of him – it was hidden behind a newspaper.

I looked at my watch and decided that it was time for lunch. From my bag I took a loaf of bread, an orange, a banana, a bottle of wine, and – most important – a corkscrew.

I had eaten the bread and polished off most of the wine when I realized that the black-suited man was watching me through a small hole he had torn in his newspaper. This was disturbing. Was he a private detective? A plain-clothes policeman? A member of the KGB? I tried to work out what I might have done wrong and began to edge towards the door. It was then that he stole my banana.

I was trying to decide between recovering my property and making a dash for freedom, when the end of the banana pushed its way through the hole in the newspaper.

Curiosity, they say, killed the cat – and that had eight lives more than I do. But the whole thing was just too bizarre. "What," I said, "do you think you're doing with my lunch?"

The newspaper was lowered and a thin face with spectacles and long fair hair appeared. The man wore a black cloak and had what looked like a wooden chain round his neck; he carried a strange stick carved into spirals with a cleft at the top like a devil's horns. Apart from the spectacles he looked like an Old Testament prophet. "I was pushing your banana through this hole," he said. "I bet you £5 you can't push your orange through it too."

"Of course I can't! The hole's too small!"

He smiled. "So if I push your orange through the hole, you owe me £5?"

"Provided you don't tear the paper, or cut up the orange, certainly!" I snapped.

He picked up the orange and held it close to the hole in one hand. With the other he stuck a finger through the hole and gave the fruit a push. "There!" he said. "As promised, I have just pushed your orange through the hole!"

I should have known better than to accept a sucker bet. I opened my wallet and transferred part of its contents into his palm. He produced a second bottle of wine and borrowed my corkscrew.

It was a curious way to start a friendship, but that's how I first met Matthew Morrison Maddox. Maddox is a professional magician, and he usually specializes in tricks that have mathematical features to them. Hence his stage name, which he normally writes "Matt M. Maddox". He told me that his stick, which he had carved himself, was called a thumb-stick: you grasp it by putting your thumb between the two tiny horns.

He was dismissive of magicians who used large pieces of complicated equipment to cut ladies in half or make elephants disappear from glass cages. The best tricks, he told me, were those that used only the simplest of apparatus. "For instance, take these corks," he said. "See how I hold them between the thumb and first finger of each hand, like *this* (figure 7.1(a)). Now, using the finger and thumb of my left hand I pick up the two ends of the cork in my right hand; and using the finger and thumb of my right hand I simultaneously pick up the two ends of the cork in my left hand." He put his hands together, pulled them apart, and each hand held a separate cork (figure 7.1(b)). "Now you do it."

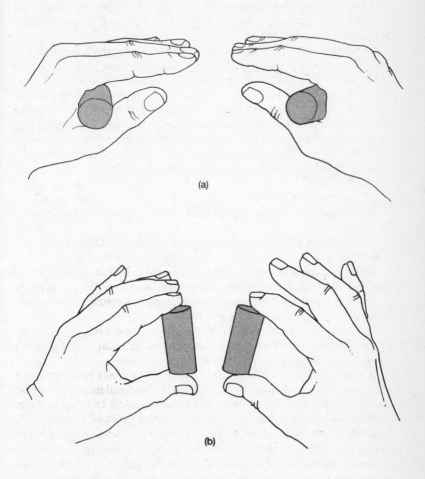

7.1 *A trick with corks. (a) Starting position. (b) Finishing position.*

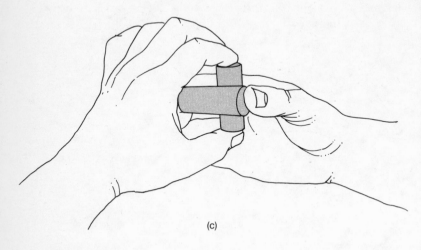

(c)

7.1   A trick with corks. (c) How to get stuck.

"Easy," I said, grasping corks between digits. But when I tried to separate my hands, the two corks wouldn't pull apart (figure 7.1(c)). He showed me the trick several times, but no matter how closely I watched, I couldn't see how he did it. "Whenever I try to do it, they're linked together," I complained. "How do you manage to unlink them? I don't see it." He just sat there, looking inscrutable.

Stop fooling around, I told myself, *and try a bit of logical thought.* I started drawing diagrams to show how the corks and my fingers got tangled up.

Reduced to the bare essentials, the arrangement that I was getting was like figure 7.2(a). Each hand+cork system forms a closed loop, and the two hand+cork systems are linked together. "It's topology, isn't it?" I said. He nodded, but still kept his silence. "I have to move my fingers so that I don't form a link . . . Let's see . . . How about *this*?" I drew figure 7.2(b). "No, silly, that's still linked, isn't it? Maybe figure 7.2(c)? No, same problem . . . Aha! If I do it *this* way (figure 7.2(d)) then the two loops just pull apart! Now, let me try with the fingers . . . Crikey, it's complicated, isn't it? Pardon me, I think I've dislocated my thumb."

So then he took pity on me. "You're right about the linking," he said. "Basically it's a topological problem. But in topology you can stretch things as much as you like, or bend them in any conceivable direction. That's not true of your thumbs, as you've just discovered. So you have to

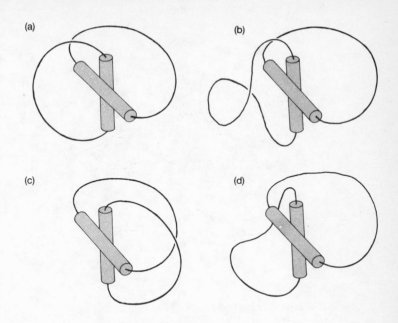

7.2 *Abstract structure of the trick. Only (d) works.*

find a way to get the fingers and thumbs into the right places without doing any damage." And then he showed me how you have to get your thumbs properly aligned, and then reach *round* with your fingers (figure 7.3).

I practised the moves a few times until they came naturally. "Hey! That's a really good party piece!"

"More of a *parity* piece, actually," he said. "You're onto a very basic distinction. Take a look at this chain." He passed me the chain from around his neck. It *was* wood, and each link was perfect and unbroken.

"How did you join the links?" I asked.

"You think I carved wooden links separately and then fitted them all together, the way a jeweller makes a metal chain? Can't be done, my friend. No, I started with a solid piece of wood and carved the chain already linked. Which leads us to a fundamental mathematical principle.

"If I take two circles in space, they can either be *linked* (figure 7.4(a)) or *unlinked* (figure 7.4(b)). If they're unlinked, I can pull them completely apart (figure 7.4(c)). The only difference in the first two pictures is the

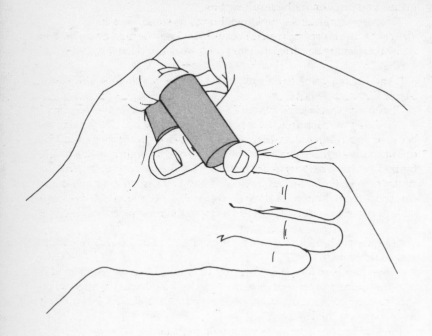

7.3 *How to separate the corks.*

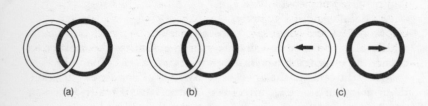

(a)   (b)   (c)

7.4 *Two circles can be linked (a) or unlinked (b), and only unlinked circles can be separated (c).*

way one crossing changes from over to under – but that tiny change means you can't undo the resulting link."

"I've seen magicians unlink metal rings on stage," I said.

"Yes. And you're impressed because you know it can't be done. You know there has to be a trick, but you can't work out what it is, can you?"

"No."

"And I'm not going to tell you, either. I value my membership of the Magic Circle too much."

"I've always wondered if there's some special way to move the rings."

"No, they're trick rings, I'll give that much away. They must be, because you can *prove* mathematically that two linked rings can't be unlinked. It's impossible to move the circles in figure 7.4(a) to get figure 7.4(c) by a *continuous deformation*. That means you can't break the circles or pass them through each other: all you can do is stretch them, compress them, and bend them. Incidentally, even after they've been distorted in shape, a topologist still refers to them as circles – and so will I."

"I've always been sure it couldn't be done," I said, "but I didn't know you could *prove* it."

"Then you're in for a treat, because I'm going to show you. The idea is to find some property of the linked circles that doesn't change when they're continuously deformed, but which the unlinked loops don't have. Can you think of one?"

I thought hard. "Um . . . Being linked together?"

"Brilliant, but not very helpful. I may be asking you to reason about circles, but I won't accept circular reasoning! I want something more definite than that. Perhaps it will help if you look at an example." He drew figures 7.5(a) and (b). "Do you agree that you can deform these two links continuously into each other?"

"Yes . . . You just push a little loop out from the black circle and feed it over the top of the white one."

"Excellent! And if I give you any pair of circles, all tangled up – coloured black and white, say, in case we need to keep track of which is which – is there any other way you can use a continuous deformation to change the way one of them overlaps the other?"

"Well, you can *undo* the move we've just been talking about, but I can't think of anything . . . Oh, you could poke a little loop of the black circle *under* the white one."

"Good."

"Or undo that move too."

"Excellent. So there are just four types of move which will change the way one circle crosses the other. Let's call them *basic moves*. We either poke a loop across, or pull it back (figure 7.6). As far as the way the circles

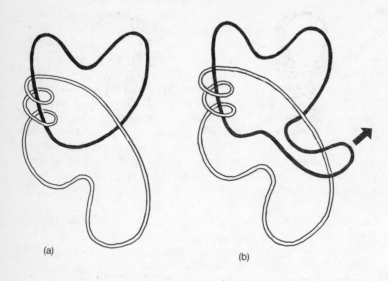

(a)   (b)

7.5 *Two links that can be deformed into each other.*

overlap each other is concerned, any continuous deformation just produces a series of basic moves, OK?"

"OK."

"Now, I want you to think about the number of crossings. How does a basic move change that?"

"It adds two. Or subtracts two. Oh! I see! If the number of crossings is even, it has to remain even! Or if it's odd, it has to remain odd!" I was getting excited now. "For two unlinked loops, like figure 7.4(c), there are no crossings, and zero is an even number! But for two linked loops, there are *two* crossings, and . . . Oh, bother!"

"Two is an even number as well."

"Yes. That's a pity. It doesn't work. If it had been an *odd* number for a pair of linked circles, that would have done it! Because you can't change an odd number into zero by adding or subtracting twos; it has to *stay* odd."

He nodded. "You're nearly there. Think about the number of crossings where the black circle goes *over* the white one. Ignore those where it goes *under*."

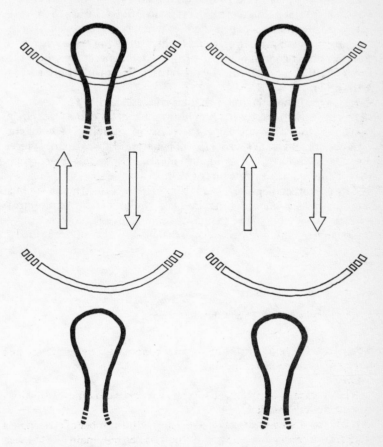

7.6 *The four basic moves that change the way two circles overlap.*

"Let's see, then . . . Well, if I do a basic move that pokes a loop of black circle over the white one, I add two to that number. If I undo the move, I subtract two. If I do the other type of basic move, poking a loop under the white circle . . . the number stays the same. Great! So the number of times the black circle goes over the top of the white one either stays the same or changes by two. If it starts out even, it stays even; if it starts out odd, it stays odd . . . Yes! And for figure 7.4(c) it's zero, which is even; but for figure 7.4(b) it's *one*, and that's odd!"

"And that's the proof," said Maddox. "It's one of the topologist's favourite tricks – playing off even against odd. You'd be surprised how powerful an idea it is. But there's an even more powerful idea involved. What you've actually done is define a *topological invariant*. That's something that you can calculate, and which stays the same when you deform an object continuously. If you take two objects for which the invariant is different, then you obviously can't deform one continuously into the other."

"Or else the invariant would be the same . . . And it isn't. Clever!"

"Here the invariant is the *parity* of the number of times the black circle goes over the top of the white one. The parity of a number is whether or not it is even or odd. That's what I meant when I said it was a parity piece. Not so easy to think of, is it? And most invariants in topology are much harder to find than our *parity invariant*."

He drew another diagram (figure 7.7). "Here's a case in point. For this pair of circles, the number of times the black circle crosses over the top of the white one is four, so the parity invariant is 'even'. So is the parity invariant for two unlinked circles. Does that mean we can undo the link?"

7.7   *A link with even parity invariant . . . Can it be undone?*

"Yes, of course. Let me try . . . Hmm, it's kind of hard to see . . . I don't know."

"Let me put it another way. If you see a black bird and a yellow bird, they can't be the same species. But if you see two black birds, does that mean they necessarily belong to the same species?"

"No. They might be a crow and a raven."

"Right. Colour is an invariant of bird species – well, provided you ignore things like budgerigars which can come in several colours. Birds of different colours are necessarily of different species, but birds of the same colour *might not* be the same species. There again, they might: colour isn't a good enough invariant to decide the question.

"In the same way, if two links have different invariants, they have to be topologically different; but that doesn't mean that if they have the same invariant, they are topologically the same. So even parity *doesn't* mean we can undo the link; what it means is that the parity invariant isn't good enough to prove that we can't."

I found that harder to follow for links than it had seemed to be for birds, and I said so.

"All right, let me give you a more mathematical analogy. Suppose I 'defourm' numbers by adding or subtracting 4 repeatedly. I want to know whether I can defourm 3 into 5. Now the parity of the number is again an invariant: adding or subtracting 4 keeps odd numbers odd and even numbers even. And 3 and 5 have the same parity. So *can* I defourm 3 into 5?"

"Well . . . 3+4 is 7, that's no good . . . 7+4 is 11; then I take away 4 again to get – bother, it's 7 again . . ."

"Exactly, and in fact it can't be done. But the parity isn't a powerful enough invariant to prove that. There's a better one. I can assign to each number an invariant equal to either 0, 1, 2, or 3; namely, its remainder when divided by 4. Let me call that the *quadruplexity*, by analogy with 'parity'."

"Ah," I said sagely. "Arithmetic modulo 4."

"Quite. Now, you can check that the quadruplexity is also an invariant under 'defourmation'. The number 3 has quadruplexity 3, but 5 has quadruplexity 1. So they're definitely *not* defourmable into each other. Do you see what I've done?"

"I'm still confused."

"The original invariant, parity, wasn't good enough to distinguish 3 and 5. They have the *same* parity. But by finding a *better* invariant, namely quadruplexity, I can show that in fact they are not defourmable into each other."

"Got it."

"Which means you have to be careful with invariants. You can use them to tell the difference between things, when they work; but when they don't, that doesn't mean the things are the same."

"I see. Is there anything better than quadruplexity? What about *octuplexity*, remainder on division by 8, or *hexadecuplexity* . . . Oh, no, that won't . . ."

"Calm down, you're getting over-excited. No, quadruplexity is a *complete* invariant. It's so good that two numbers can be defourmed into each other *if and only if* they have the same quadruplexity invariant."

"Yes, but octuplexity will be a better invariant if I *transleight* numbers by adding or subtracting 8 repeatedly . . ."

"Concentrate on the problem, you're entering a manic phase."

I did. Eventually light began to dawn. "You're telling me that there's a better invariant than the parity invariant for links, and it tells me that I can't change figure 7. 7 into figure 7.4(c). Let me guess: quadruplexity of the number of crossings?"

"No, not as easy as that. What you need is one of the first topological invariants ever discovered, and it's called the *linking number*. The way you find it is that you imagine a membrane stretched across the white circle, like a paper hoop. You put an arrow on the black circle and see where it passes through the membrane, following the direction of the arrow along the circle. If it cuts through from back to front you count +1, if it cuts from front to back you count −1. Then you just add those +1's and −1's at all places where the black circle cuts through the membrane.

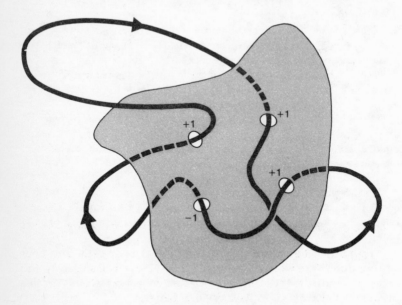

7.8 *Calculating the linking number.*

"For instance, in figure 7. 8 with the arrow as shown, the black circle passes through the membrane three times from back to front and one time from front to back, so the linking number is $1 + 1 + 1 - 1 = 2$."

"That's interesting," I said. "It sort of counts how many times the black circle wraps round the white one."

"Good. Now the linking number is an invariant. Well, you have to choose how to align the arrow, and if you reverse the arrow the linking number changes sign, so that *after calculating it* you should change the sign to + for safety. But that's easily taken care of. The question is, do you see *why* it's an invariant?"

"No," I told him. I'm a great believer in frank and meaningful discussions.

"It's basically the same trick again. The only way to change the linking number is to create or destroy *two* cutting points by pulling a loop through figure 7.9. But when you look at the directions . . ."

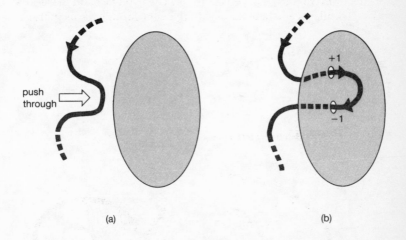

(a)   (b)

7.9   *The linking number is unchanged if a pair of cutting points is created or destroyed.*

"You see that the two points count as $+1$ and -1, because the directions are opposite!" The words came out in a rush, I saw it all so clearly. "So when you add up to get the linking number they just cancel out! And that means the linking number itself doesn't change!"

"Exactly. Isn't it amazing, the lengths that you have to go to to prove something so 'obvious'? But it isn't really obvious at all. Experience tells us it's true, but it doesn't give logical reasons why it *has* to be true."

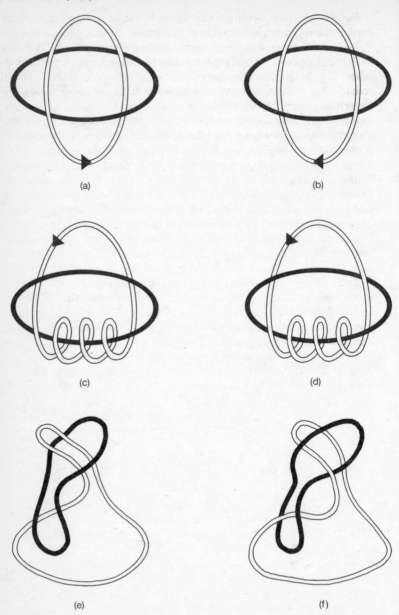

7.10   *Six links. Two are deformable into each other. Which?*

Figure 7.10 shows six links. Can you calculate their linking numbers? Two of them can be deformed into each other: which?

I asked whether the linking number is a *complete* invariant. If two links have the same linking number, can you deform one continuously into the other? Matt M. Maddox pointed out that a particular case might be worth considering: "Can you find two circles, with linking number zero, which nevertheless can't be unlinked?"

If you can, then the linking number isn't a complete invariant. In fact, it's not. Indeed, finding a complete invariant for links is still an unsolved problem. However, there are several new link invariants around, one of which was discovered simultaneously by five independent groups of mathematicians, and one of them might *just* turn out to be complete, even though nobody really believes it will . . . But that's another story, well told by Lickorish and Millett: see "Further Reading". Anyway, can you answer Matt M. Maddox's question about linking number zero?

I couldn't, I'll tell you. After twenty minutes I wiped my brow with a handkerchief, and drained the last of the wine. "All that mathematics from two corks! I wonder how much you'd find in a bottle!"

"Many have found poetry in a wine bottle," said Maddox. "Not much mathematics, though – you need to keep a clear head for that. Mind you: 'A loaf of bread, a glass of wine, and thou . . .' Omar Khayyam was a pretty good mathematician – he found a geometric method for solving cubic equations. Which reminds me – there's an American manufacturer of alcoholic beverages who puts some rather special links on its label (figure 7.11). They're known as *Borromean rings*, because they come from the coat of arms of the Borromeo family."

7.11   *Borromean rings. One cut separates the lot.*

I stared at the label. "Doesn't look very special to me . . . Just three circles, all linked up."

"Hmm. If you have three circles, all linked up, how many do you have to cut to get three separate circles?"

"Two."

"Why?"

"Each cut separates off one circle."

He produced three loops of string, arranged just like the Borromean rings, and a pair of scissors. He jiggled the rings up and down to show they were all joined together. "Cut *one* loop," he said. I did.

All three rings fell to the table, separately.

"Ah, you cheated. If you have three circles in a row, you can cut the middle one, and they all fall apart."

He handed me another set of Borromean rings. "Fine. You choose which one to cut."

"Well, I want the one on the end . . . funny, there isn't an end . . ."

"It's symmetrical," he pointed out. "Whichever ring you cut, all three will separate."

"Huh," I said. "Well, I bet you can't do that with *four* rings!"

He looked at me, once more inscrutable.

"Do you want to bet £5?" he said.

7.12 *The Whitehead link has linking number zero but cannot be separated.*

ANSWERS

The six linking numbers for figure 7.10 are $a=1$, $b=2$, $c=4$, $d=5$, $e=0$, $f=2$. So links b and f are the only ones that could possibly be deformable into each other, and indeed a little experiment shows that they are.

Two circles with linking number zero, which are nevertheless linked, are shown in figure 7.12. This is known as the *Whitehead link*.

Not only can you find a link of four circles which falls completely apart when only one is cut: you can find such a link with n circles for any n. Figure 7.13 shows how this is done for seven circles: the general pattern should be obvious. Elegant answers for four and five rings, different from mine, were shown to me by C. van de Walle of Evreux (figure 7.14).

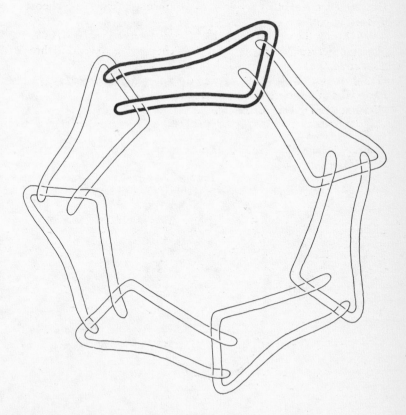

7.13   *Generalized Borromean rings. Remove any ring (e.g. the shaded one) and the rest slide apart.*

7.14   Elegant solutions for four and five rings, due to C. van de Walle.

FURTHER READING

J. H. Conway, "An Enumeration of Knots and Links, and Some of their Algebraic Properties", *Computational Problems in Abstract Algebra*, ed. J. Leech (Oxford: Pergamon Press, 1969), pp. 329–58

H. M. Cundy and A. P. Rollett, *Mathematical Models* (Oxford: Oxford University Press, 1961)

G. Kolata, "Solving Knotty Problems in Math and Biology", *Science*, 231 (28 March 1986), pp. 1506–8

W. B. R. Lickorish and K. C. Millett, "The New Polynomial Invariants of Knots and Links", *Mathematics Magazine*, 61 (February 1988), pp. 3–23

D. Rolfsen, *Knots and Links* (Berkeley: Publish or Perish, 1976)

8

Close Encounters of the Fermat Kind

I've told you about my garden, behind the raspberry canes. There's a space-time warp there. It leads, among other things, to the planet Ombilicus, a billion light-years from Earth in the general direction of Orion's right eyeball. Where – and when – you end up if you pass through it seems to depend upon what you're thinking about when you enter it. I've been doing a few experiments to see whether I can fine-tune the position and timing. I haven't quite got the hang of the thing yet – last week I narrowly avoided emerging in ancient Jericho just as the walls collapsed – but I'm getting there.

Yesterday I decided to drop in on a mathematician whom I greatly admire, namely Pierre de Fermat, who was born in 1601 and died in 1665. He lived in Toulouse, and was a lawyer. He's best known, of course, for the one theorem that he didn't prove (or did he?), namely Fermat's Last Theorem.

You've probably heard the story. Fermat owned a copy of the *Arithmetica* of Diophantus, an early algebra book. Equations that must be solved in integers (whole numbers) are called *Diophantine equations* in his honour. At one place Diophantus explains how to find right-angled triangles whose sides are all integers. By Pythagoras' theorem such triangles have sides (a,b,c) such that $a^2 + b^2 = c^2$. There are infinitely many integer solutions of this equation, such as $(3,4,5)$ and $(5,12,13)$. Anyway, Fermat started thinking about sums of two perfect squares being perfect squares, and wondered whether the same sort of thing can be done for cubes, or biquadrates (fourth powers), or whatever. Can two perfect cubes add up to another perfect cube?

He decided it was impossible, and wrote as much in the margin of his book. "It is impossible to separate a cube into two cubes, or a biquadrate into two biquadrates, or generally any power except a square into two powers with the same exponent. I have discovered a truly marvellous proof of this, which however the margin is not large enough to contain."

To this day, nobody has been able to reconstruct the missing proof; neither have they found any example to show that Fermat's Last Theorem is false. The problem is notorious. A huge prize was once on offer for the answer, but its value was wiped out by inflation, and it has now been replaced by something financially more modest. It's named the "last theorem" because it is the last remaining unsolved riddle from those that Fermat posed for his successors. Many have tried, but none have yet succeeded.

Modern mathematicians find it hard to believe that Fermat knew something that they don't – although personally it wouldn't surprise me in the least – and they tend to assume that if Fermat thought he had a proof, there must have been a mistake in it. Hundreds, maybe thousands of plausible but fallacious proofs have been invented since, and it's certainly *possible* that Fermat blundered into one of these traps. But Fermat was one smart cookie: maybe he was right.

Or maybe not. Which? Most people consider it impossible to resolve that question: there's no evidence either way. But then, most people don't have a space-time warp behind the raspberry canes. My plan was to go back to Fermat's time and ask him myself. Thinking that he might be curious about the fate of his conjecture, I collected together some information on its current status to take with me.

Then I fixed my mind firmly on Fermat's Last Theorem, and walked into the space-time warp.

It worked, I must say, like a charm. I came out in a comfortable room, full of antique furniture, with a log fire blazing in the hearth. A bewigged figure was sitting at a desk, quill pen in hand, writing in a notebook. I cleared my throat, to catch his attention, and he turned.

"Whence came you?" he cried, in some alarm, leaping to his feet and brandishing his quill pen like a club. "Art thou a thief, come to steal my valuables?"

"No, Monsieur de Fermat," I replied, keeping pretty close to the warp just in case he was armed with a pistol or attempted to knock me down with his feather. "I am an admirer from the distant future."

Fermat considered this. He looked at my clothes – old jeans and a red sweatshirt bearing the legend "Phi Slama Jama: Texas's Tallest Fraternity". (It refers to the University of Houston basketball team. I spent a year there once.) "Mayhap thou art right," he said. "Thy dress is outlandish and not from this time, and thy speech is atrocious. But then, thou'rt English, so that proveth little. Yet there is a strangeness to thy accent, even so." Then, ever the mathematician, he darted a penetrating question. "Canst thou *prove* thy claim?"

I had anticipated this, and I'd brought a programmable calculator with me. Ten minutes showing him how to generate Fibonacci numbers or solve cubic equations to ten decimal places, and he was convinced.

"Why art thou here, Traveller in Time?"

I explained that in the far future he was an extremely famous mathematician. This greatly surprised him. "But 'tis merely a pastime of mine, a small conceit to while away the hours!"

I waved him to silence and told him not to be so modest. "Pierre, I've come to ask you about your Last Theorem," I said.

"My what?"

"You won't know it by that name, of course. The theorem that it is impossible to resolve a cube into two cubes, or a biquadrate into two biquadrates, or . . ."

He looked baffled and pulled at his wig. "That is an idea of great interest. I have never contemplated the question: it hath a certain *je ne sais quoi* . . . But I know not what. I will jot it down in my copy of Diophantus' *Arithmetica* . . ."

That's the problem with time travel – you never know what paradoxes you may cause. Here was I, coming to ask Fermat about his theorem, and now he'd learned about the problem from *me*!

He looked at the clock on the wall and leaped to his feet. "Forgive me, but I have an urgent appointment in court. Perhaps thou wouldst pay thy respects again, some time in the future? A week from now?" And he was gone. I took one final look at his study, and backed out through the warp, wondering what I'd done. Would the universe still exist?

You'll be glad to hear it did. I convinced myself that by going back in time and putting the idea into Fermat's head I'd probably *saved* the universe from paradoxical dissolution. Well, if Fermat wasn't going to think of it by himself, *someone* had to, otherwise history would have been changed . . .

The great thing with space-time warps is that you don't have to hang about. Setting my mental clock for a week later than my first attempt, I turned on my heels . . .

Fermat was expecting me. "Good day, Traveller in Time! Thou hast posed a pretty riddle, I warrant! Seven days and nights hath it troubled my mind. 'The equation $x^n + y^n = z^n$ is impossible in integers, when *n* is 3 or more.' I can find no instances where thy conjecture . . ."

"No, no, *your* conjecture! Otherwise the universe may dissolve!"

"Very well, where *my* conjecture faileth. I *have* found several instances where it faileth but by a hair's breadth. Thus,

$$9^3 + 10^3 = 1729, \ 12^3 = 1728,$$

so that the equation

$$x^3 + y^3 = z^3 + 1$$

hath a solution. For

$$x^3 + y^3 = z^3 - 1$$

I have also found solutions." *Can you find one? Or any other "Fermat near misses": solutions to these two equations? And what about* $x^4 + y^4 = z^4 \pm 1$?
"And I have found innumerable other curious relationships between powers," he said (box 8.1). "But," he went on, "I can find no cube that is exactly divisible into two cubes. I can find a cube that divideth into two squares, and a square that divideth into two cubes." (*Can you?*) "I *have* found a proof of impossibility in one case, namely that of biquadrates." *That was quick*, I thought. "The idea is an amusing one, I will admit. I call it . . ."

"The method of infinite descent, yes."

"Thou knowest it?"

"I told you; in the future you're famous. And so is your proof for fourth powers."

He shook his head in wonderment. "But I am a mere amateur."

"Blaise Pascal called you 'the greatest mathematician in all Europe'."

"Pascal is a flatterer; he's always after something." He sighed. "The future . . . I would give much to know what new wonders of mathematics will be found . . . And to possess a 'programmable calculator' such as thine."

Box 8.1

$$133^4 + 134^4 = 158^4 + 59^4$$
$$1^4 + 8^4 + 12^4 + 32^4 + 64^4 = 65^4$$
$$4^4 + 6^4 + 8^4 + 9^4 + 14^4 = 15^4$$
$$30^4 + 120^4 + 272^4 + 315^4 = 353^4$$
$$1^4 + 2^4 + 9^4 = 3^4 + 7^4 + 8^4$$
$$5^4 + 6^4 + 11^4 = 1^4 + 9^4 + 10^4$$
$$8^4 + 9^4 + 17^4 = 3^4 + 13^4 + 16^4$$
$$7^4 + 28^4 = 3^4 + 20^4 + 26^4$$
$$51^4 + 76^4 = 5^4 + 42^4 + 78^4$$
$$4^5 + 5^5 + 6^5 + 7^5 + 9^5 + 11^5 = 12^5$$
$$49^5 + 75^5 + 107^5 = 39^5 + 92^5 + 100^5$$
$$3^6 + 19^6 + 22^6 = 10^6 + 15^6 + 23^6$$

"Pierre, I'd love to give you one, but I'm afraid it might cause a time paradox, so I daren't. But I can fill you in on what's happened to your Last Theorem."

And I told him how various of his successors had proved special cases (table 8.1), and that his conjecture was known to be true for all powers up to and including the 125,000th. He opened his copy of Diophantus, took up his pen, and began writing rapidly in what I perceived to be an *enormous* margin, taking down everything I said in some legal shorthand . . . It made me feel very nervous, because history records no such annotations, but it seemed impolite to stop.

"The most dramatic new result in my time, " I told him, "is the proof of the Mordell conjecture in 1983, by a young German called Gerd Faltings. Mordell conjectured that for a whole class of Diophantine equations, including yours, the number of solutions is finite. Faltings found an extremely advanced and difficult proof. So for all $n \geq 3$, if there are any exceptions to your Last Theorem, there are at most a finite number of them."

Not everyone appreciates that, even if you don't know the exceptions exactly, it's a great step forward to know that their number is finite. It's important to know that, because then you can hope to set limits on their sizes, after which in principle a trial-and-error check would finish the problem altogether. In practice the limits are often too large for this to work, but by being more clever you can still hope to get somewhere.

Indeed, by using Faltings's result, D. R. Heath-Brown proved in 1987 that Fermat's Last Theorem is true for "almost all" exponents n. That is, as n tends to infinity the proportion of values for which the Last Theorem is true tends to 100 per cent. That's a strong result: instead of saying that for each n there may be finitely many solutions, it says that for all but a very rare set of exponents n there are *no* solutions. As I explained this, Fermat again scribbled copious notes in the wide margins of his *Arithmetica*, and I squirmed.

Table 8.1   Milestones in Fermat's Last Theorem

Date	Discoverer	$x^n + y^n = z^n$ impossible ...
c.1640	Pierre de Fermat	$n = 3$
c.1640	Pierre de Fermat	$n = 4$
1738	Leonhard Euler	$n = 3$ (independently)
1738	Leonhard Euler	$n = 4$ (independently)
c.1815	Sophie Germain	if $n \geq 3$, both n and $2n+1$ are prime, and n does not divide xyz
1828	Peter Lejeune Dirichlet	$n = 5$
1830	Adrien-Marie Legendre	$n = 5$ (independently)
1832	Peter Lejeune Dirichlet	$n = 14$
1859	Ernst Eduard Kummer	n a "regular" prime: in particular $n \leq 100$ except 37, 59, 67
1893	Dimitri Mirimanoff	$n = 37$
1905	Dimitri Mirimanoff	$n \leq 257$
1909	A. Wieferich	n an odd prime not dividing xyz with n^2 not dividing $2^{n-1} - 1$ (the second condition holds for all $n < 3 \times 10^9$ except 1093 and 3511)
1922	Leo Mordell	with finitely many exceptions for any $n \geq 3$ provided the "Mordell conjecture" is true
1978	S. S. Wagstaff	$n \leq 125,000$
1983	Gerd Faltings	with finitely many exceptions for any $n \geq 3$
1987	D. R. Heath-Brown	for "almost all" n

"I have proved a small number of results on the finitude of solutions myself," he said modestly. "My favourite is that the only cube exceeding a square by 2 is $3^3 = 5^2 + 2$."

"In that case you'll like W. Ljunggren's theorem that $1 = 1^2$ and $57121 = 239^2$ are the only squares which, when increased by one and halved, yield fourth powers."

He looked fascinated. But then his expression changed. "But there must be some error in thy description of this work of Monsieur Faltings. If $x^n + y^n = z^n$ then for any constant k we obtain $(kx)^n + (ky)^n = (kz)^n$. Thus one solution generateth an infinity."

"That's true," I said. "I meant infinitely many solutions without a common factor."

"Ah."

"But actually, that's not the way to think of it. The way Mordell and Faltings thought about it is to notice that the equation $x^n + y^n = z^n$ is equivalent to $(\frac{x}{z})^n + (\frac{y}{z})^n = 1$. Putting $\frac{x}{z} = X$ and $\frac{y}{z} = Y$ you see that solving $x^n + y^n = z^n$ in *integers* is equivalent to solving $X^n + Y^n = 1$ in *rational* numbers."

"Yes, I am aware of that. Much of my work has been about solutions of equations in rational numbers."

"Multiplying x, y, and z by a constant k does not change X or Y. So the number of rational solutions to $X^n + Y^n = 1$ is finite, without any quibbles. From that point of view," I continued, "your Last Theorem is rather curious. The equation $X^n + Y^n = 1$ defines a curve in the (X,Y) coordinate plane which nowadays we call the *Fermat curve* of degree n." When n is even, Fermat curves are like squarish ovals; when n is odd they extend to infinity (figure 8.1). "The Last Theorem says that even though points (X,Y) with both coordinates rational are dotted densely throughout the plane, the Fermat curve winds its way between such *rational points*, never hitting any of them."

"But that proveth little," he said. "For indeed there are many such curves. Thus the straight line $Y = X + \sqrt{2}$ cannot meet any rational point. If it were to do so, then $\sqrt{2} = Y - X$ would be rational."

"Yes, but the coefficients in the Fermat equation are themselves rational, whereas $\sqrt{2}$ is not."

"But a simple trick converteth the line $Y - X = \sqrt{2}$ into $(Y - X)^2 = 2$, that is, $X^2 + Y^2 - 2XY = 2$, and now no irrationals appear in the equation." (Figure 8.2)

"True," I said. "It shows how careful you have to be with this kind of thing. At any rate, Faltings' result is that each Fermat curve can hit only a finite number of rational points."

Fermat thumbed the pages of his book. "Thy question hath set me to muse also on related matters," he said. "For instance, if it be not possible for *two* cubes to sum to a cube, might it be possible for *three*? And of course it is; in fact $3^3 + 4^3 + 5^3 = 6^3$. And that leadeth me to conjecture that for all n it is possible for n nth powers to add to an nth power, but not for $n-1$ to do so."

He scribbled excitedly in the margins, to my growing horror. "But that's Euler's conjecture!" I yelled. "It came years after yours! Please,

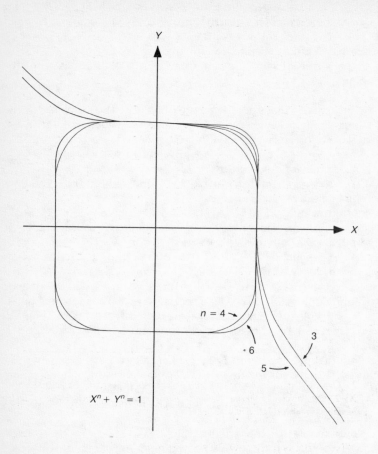

8.1　*Fermat curves of various degrees. If the Last Theorem is true, no Fermat curve contains any point with both coordinates rational numbers.*

whatever you do, don't put it into print! The paradoxes would be too dreadful!"

"Unless," Fermat mused, "the printed versions failed to survive to thy time. Then thou wouldst not know that I had anticipated Monsieur Euler."

"Maybe." I wasn't happy. *What about all those marginal annotations he was making? And why had he complained about the margins of his book, which were absolutely vast?* I tried to distract him. "In my time it has been proved that Euler's conjecture is false."

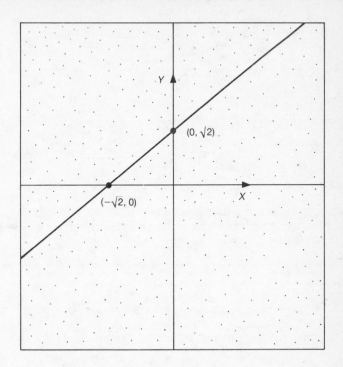

8.2 *Curves that pass through no rational points can occur. The curve* $Y^2 - 2XY$ *+ X^2–2 = 0, which is just the line $Y = X + \sqrt{2}$, is an example. The rational points, illustrated schematically by dots, are dense in the plane.*

"Now that is truly news of some import."

I explained that in 1966 L. J. Lander and T. R. Parkin had found the first (and until 1988 the only) counterexample to Euler's conjecture: four fifth powers whose sum was again a fifth power (box 8.2).

"How did they find it?" he asked, as he copied the numbers into the margin. I groaned inwardly. By now he had filled most of the margins of his book.

"By exhaustive computer search."

"Computer?"

"A giant programmable calculator."

"Oh. I had hoped there might be some interesting mathematics involved."

"There is! In 1988 Noam Elkies of Harvard University found another

---

Box 8.2

$$27^5 = 14348907$$

$$84^5 = 4182119424$$

$$110^5 = 16105100000$$

$$135^5 = 41615795893$$

$$144^5 = 61917364224$$

---

counterexample: three fourth powers whose sum is a fourth power (box 8.3). And that did involve some genuine mathematics, not just a computer search."

Box 8.3

$$2682440^4 = 51774995082902409832960000$$

$$15365639^4 = 55744561387133523724209779041$$

$$18796760^4 = 124833740909952854954805760000$$

$$20615673^4 = 180630077292169281088848499041$$

"Tell me more."

"Well, Elkies started the same way Faltings did: instead of looking for integer solutions to the equation $x^4 + y^4 + z^4 = w^4$ he divided out by w^4 and looked at the surface $r^4 + s^4 + t^4 = 1$ in coordinates (r,s,t). It's kind of a cross between an ellipsoid and a cube (figure 8.3). An integer solution to $x^4 + y^4 + z^4 = w^4$ leads to a rational solution $r = \frac{x}{w}$, $s = \frac{y}{w}$, $t = \frac{z}{w}$ of $r^4 + s^4 + t^4 = 1$. Conversely, given a rational solution of $r^4 + s^4 + t^4 = 1$, you can assume that r,s,t all have the same denominator w by putting them over a common denominator, and that leads directly to a solution to $x^4 + y^4 + z^4 = w^4$."

"Yes, yes, that is clear." Well, to people like Fermat it doubtless is.

"A Russian mathematician, V. A. Demjanenko, found a rather complicated condition for a point (r,s,t) to lie on the related surface

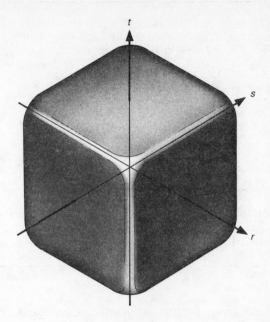

8.3 *Elkies' surface* $r^4+s^4+t^4 = 1$. *Unlike the Fermat curve* $X^4+Y^4 = 1$, *points with all three coordinates* (r,s,t) *rational are dense on this surface.*

Box 8.4

Demjanenko showed that $r^4 + s^4 + t^2 = 1$ if and only if there exist $x, y,$ and u such that

$$r = x + y$$
$$s = x - y$$
$$\left(u^2 + 2\right)y^2 = -\left(3u^2 - 8u + 6\right)x^2 - 2\left(u^2 - 2\right)x - 2u$$
$$\left(u^2 + 2\right)t = 4\left(u^2 - 2\right)x^2 + 8ux + \left(2 - u^2\right)$$

$r^4 + s^4 + t^4 = 1$ (box 8.4). To solve the problem it is enough to show that t can be made a square. A series of simplifications shows that this can be done provided the equation

$$Y^2 = -31790X^4 + 36941X^3 - 56158X^2 + 28849X + 22030$$

has a rational solution. There is an extensive theory of such equations, known as *elliptic curves*. In particular there are conditions under which *no* solution can exist. These conditions did not hold in this case, which showed that such a solution *might* possibly exist. At this stage Elkies tried a computer search, and found the solution

$$(X, Y) = \left(-\tfrac{31}{467}, \tfrac{30731278}{467^2}\right).$$

From this he deduced the rational solution

$$(r, s, t) = \left(-\tfrac{18796760}{20615673}, \tfrac{2682440}{20615673}, \tfrac{15365639}{20615673}\right)$$

This led directly to the counterexample to Euler's conjecture for fourth powers, namely $2682440^4 + 15365639^4 + 18796760^4 = 20615673^4$."

"So even here it was necessary to use one of these 'computers'?" Fermat said, rather grimly, while his quill flickered ominously down the margin. "Do the mathematicians of thy time not use their own heads any more?"

"Most of the time. Even here Elkies used his head first, and then used the computer in a more intelligent fashion than just a trial-and-error search. Computers are tools to help mathematicians, not replacements for them."

"I see. Are there other solutions?"

"Yes. In this case infinitely many. The theory of elliptic curves provides a general procedure for constructing new rational points from old ones, using the geometry of the curve." (Figure 8.4)

"That is an ancient trick. I have seen its like in the work of Monsieur Bachet."

"It does go back a long way, although I doubt you would recognize its most general form, the theory of Abelian varieties. But I digress. By applying a variant of this construction, Elkies proved that infinitely many solutions exist. In fact he proved that there are many rational points on the surface $r^4 + s^4 + t^4 = 1$; so many that they are *dense*. That is, any patch of the surface, however tiny, must contain a rational point. The numbers get very big, though. The second solution generated by this geometric construction is . . ." (I wrote down what is in box 8.5).

"Those are impressively large numbers. And I note with some satisfaction that they are found from the first solution by pure thought, not by computer search." He started to copy down the four vast numbers, but stopped. He had run out of space.

"There's a curious twist to the tale," I said, hoping to divert his attention. *I had to grab that book!* But if I stole it, then posterity would never know that Fermat had *had* a copy of the *Arithmetica* . . . Paradox lost, paradox regained . . . Best to keep talking. "After Elkies had discovered there *was* a solution, Roger Frye of the Thinking Machines Corporation

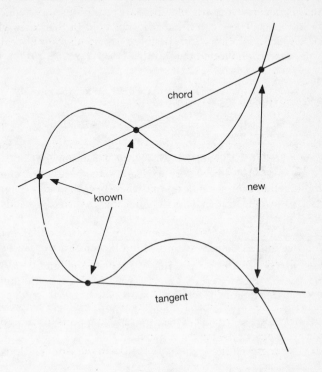

8.4 *If a straight line is drawn between two rational points on an elliptic curve, it meets the curve in a third point which must also be rational. If a tangent is drawn to an elliptic curve at a rational point, it meets the curve again in a rational point. This chord and tangent process thus produces new rational solutions from old ones, an application of geometry to number theory.*

did an exhaustive search. It took 100 hours on Connection Machines . . ."

"Thinking machines? Connection machines? Is the world of the future nothing but machines?"

"Pretty much. We live in them, eat out of them, drive around in them, fly in them . . ."

"Fly? You jest."

" . . . And use them to talk to the other side of the Earth. A Connection Machine is a supercomputer, able to perform millions of calculations every second. By using this machine-gun to swat a gnat, Frye found a smaller solution: indeed the smallest possible solution." (Box 8.6) "So the computer won in the end."

Box 8.5

Elkies' second solution. If

$x = $ 1439965710648954492268506771833175267850201426615300442218292336336633

$y = $ 4417264698994538496943597489754952845854672497179047898864124209346920

$z = $ 9033964577482532388059482429398457291004947925005743028147465732645880

$w = $ 9161781830035436847832452398267266038227002962257243662070370888722169

then

$$x^4 + y^4 + z^4 = w^4$$

"Remarkable. Millions . . . But this tale setteth me thinking again. I could refine Euler's conjecture. Thus for each n let $s(n)$ be the smallest number such that there exist $s(n)$ nth powers whose sum is an nth power. Thus $s(2) = 2$ because two squares can add up to a square. And $s(3) = 3$ because three cubes can sum to a cube, but two cannot. The conjecture of Monsieur Euler is that $s(n) = n$, but this thou sayest is false. In fact $s(4) = 3$ because Monsieur Elkies' example proveth that $s(4) \leq 3$ and my theorem for biquadrates proveth that $s(4)$ is not 2."

"And Lander and Parkin's result gives $s(5) = 4$," I said.

"Objection!" he snapped, ever the professional lawyer. "It proveth only that $s(5)$ is 4 *or less*." Fermat knew a loophole when he saw one, legal or mathematical. "By what thou hast told me, we know that $s(5)$ is 3 or more . . . *Can three fifth powers sum to a fifth power?*"

Box 8.6

$$95800^4 = 84229075969600000000$$

$$217519^4 = 2238663363846304960321$$

$$414560^4 = 29535857400192040960000$$

$$\overline{422481^4 = 31858749840007945920321}$$

"No idea. Sounds unlikely."

"Mayhap. But in any case it is an interesting problem: to calculate $s(n)$ for each n. There are some easy results, of course: for example $s(6)$ is at most 64, and in general $s(n)$ is at most 2^n."

"Because $2^6 = 1^6 + \ldots 1^6$ with sixty-four 1's, and in general $2^n = 1^n + \ldots 1^n$ with 2^n 1's."

"Exactly. I cannot believe such crude estimates are the best possible, although they have the virtue of proving that $s(n)$ is finite for every n."

(*Can you improve on these estimates?* For example, if you can find ten seventh powers whose sum is a seventh power, you will have shown that $s(7) \le 10$. That would be a lot better than Fermat's estimate of 128. You might like to experiment.)

So excited was Fermat about all this that he began pulling books off the shelf, looking for further margins to annotate. I leaped forward, grabbed the copy of Diophantus, and tore huge strips off the margins to obliterate everything he had written, throwing the sheets into the fire. I had to avoid a paradox at all costs!

It took a little time to calm Fermat down, but eventually he saw why I had acted in such a high-handed manner. He sat, staring into the flames, his expression unreadable.

Then his face . . . changed.

It was like sunrise breaking through a storm.

"A proof!" he cried. "I have a proof of the Last Theorem! 'Tis ingenious but elegant . . . Let me inscribe it in the margin! Confound thee, Time Traveller! Thou hast torn my margins to shreds, there is no longer room to write the proof! Oh, where did I put that sheaf of legal documents?"

I tiptoed from the room, back through the warp.

Time paradoxes are funny things.

ANSWERS

Fermat Near Misses
$6^3 + 8^3 = 9^3 - 1$ is an example for $x^3 + y^3 = z^3 - 1$. T. Wightman of Borehamwood, England, sent me two other examples in May 1988:

$$720^3 + 242^3 = 729^3 - 1$$
$$729^3 + 244^3 = 738^3 + 1$$

which suggest interesting patterns. In particular $729 = 3^6$, so these solve the remarkable equations

$$x^3 + y^3 = z^{18} - 1$$
$$x^{18} + y^3 = z^3 + 1.$$

In fact there are infinitely many solutions to the equations $x^3 + y^3 = z^3 \pm 1$. To find them, consider more generally the equation $x^3 + y^3 = z^3 + w^3$. It is known that the general rational solution to this is given by

$$x = k\left[1 - (a - 3b)(a^2 + 3b^2)\right]$$

$$y = k\left[(a + 3b)(a^2 + 3b^2) - 1\right]$$

$$z = k\left[(a + 3b) - (a^2 + 3b^2)^2\right]$$

$$w = k\left[(a^2 + 3b^2)^2 - (a - 3b)\right],$$

where a, b and k are rational. Choose them so that one of the variables is ± 1 and all four are integers. (For instance, let $k = 1$, $a = 3b$ to make $x = 1$.)

$x^4 + y^4 + z^4 \pm 1$

I don't know any solution to this, but I also don't know of any proof that none exist. Maybe somebody already knows the answer. Some close coincidences that I've found by trial and error are

$$7^4 + 8^4 = 9^4 - 64$$
$$21^4 + 36^4 = 37^4 - 64$$
$$11^4 + 15^4 = 16^4 - 270$$
$$37^4 + 37^4 = 44^4 + 226$$
$$53^4 + 62^4 = 69^4 - 304$$
$$167^4 + 192^4 = 215^4 + 192.$$

$x^3 + y^3 = z^2$

One solution is $1^3 + 2^3 = 3^2$. To find infinitely many, proceed as follows. Take any two integers, for example 2 and 3. Work out $2^3 + 3^3 = 8 + 27 = 35$. Multiply the original equation by 35^3 to obtain

$$(2 \times 35)^3 + (3 \times 35)^3 = 35 \times 35^3 = 35^4 = (35^2)^2.$$

That is,

$$70^3 + 105^3 = 1225^2.$$

A similar method works for any two starting numbers.

$x^2 + y^2 = z^3$

Try the same idea. Choose any two numbers, say 1 and 2. Then $1^2 + 2^2 = 5$. Multiply throughout by 5^2 to get $5^2 + 10^2 = 5^3$. Again the procedure works for any two numbers whatever.

$x^5 + y^5 + z^5 = w^5$

I have no idea whether or not this is possible!

Estimates for s(n).

This is related to a question known as *Waring's problem*. In 1770 Edward Waring stated that every whole number is a sum of at most nine cubes, nineteen fourth powers, and so on. Following his idea, number theorists defined a function $g(n)$ to be the smallest number such that every number k can be written as a sum of $g(n)$ nth powers. So Waring's conjecture is that $g(3) = 9$, $g(4) = 19$, and – by implication – that $g(n)$ is finite for all n. This conjecture was later proved by David Hilbert.

"Small" numbers k tend to require unusually large numbers of nth powers, for rather coincidental reasons. Accordingly, a more sensible function is $G(n)$, the smallest number such that *all but finitely many* numbers k can be written as a sum of $G(n)$ nth powers. A great deal of work has been done to find $G(n)$. For example in 1958 J. R. Chen proved that $G(n) \leq n(3 \log n + 5.2)$. In 1984 R. Balasubramanian and C. J. Mozzochi improved this to get

$$G(n) \leq \frac{\log (108) + 3 \log (n)}{\log \left(\frac{n}{(n-1)} \right)} - 4$$

and other mathematicians, notably R. C. Vaughan in 1986, have improved on this for various small n. The best known bounds for $G(n)$ are

n	4	5	6	7	8	9	10	11	12	13	14	15
$G(n) \leq$	19	21	31	45	62	82	102	120	135	150	166	181

Obviously our function $s(n)$ is less than or equal to $G(n)$: just take k to be a very large nth power. So the table for $G(n)$ also gives bounds on how large $s(n)$ can be. However, it is likely that $s(n)$ is *smaller* than $G(n)$ for all $n \geq 6$.

We know that this happens for $n = 4$ and 5, because $s(4) = 3$ and $s(5) \leq 4$. It is also the case for $n = 6$ and 7, as shown by examples from *The Penguin Dictionary of Curious and Interesting Numbers*, by David Wells. As he remarks, a sixth power can be the sum of seven sixth powers. The smallest example is

$$1141^6 = 74^6 + 234^6 + 402^6 + 474^6 + 702^6 + 894^6 + 1077^6.$$

Therefore $s(6) \leq 7$. For seventh powers, we have

$$102^7 = 12^7 + 35^7 + 53^7 + 58^7 + 64^7 + 83^7 + 85^7 + 90^7,$$

showing that $s(7) \leq 8$. This is the smallest seventh power that is a sum of eight seventh powers.

X. Gonze of Louvain, Belgium sent me the following discoveries:

$$12^8 = 2 \times 11^8 + 3 \times 5^8 + 4^8 + 4 \times 3^8 + 2^8 + 23 \times 1^8$$
$$5^9 = 7 \times 4^9 + 6 \times 3^9 + 19 \times 1^9$$
$$7^{11} = 5 \times 6^{11} + 3 \times 5^{11} + 4 \times 4^{11} + 39 \times 2^{11}$$

which show that $s(8) \leq 34$, $s(9) \leq 32$, and $s(11) \leq 51$.

To sum up, the best known bounds on $s(n)$ are

n	4	5	6	7	8	9	10	11	12	13	14	15
$s(n)$	3	4	7	8	34	32	102	51	135	150	166	181

The bounds for $n = 10$ and $n \geq 12$ come from $G(n)$, but the pattern of the numbers strongly suggests that these bounds are far too big.

Two other noteworthy results from Wells's book are as follows. The smallest fourth power that is the sum of five fourth powers is

$$5^4 = 2^4 + 2^4 + 3^4 + 4^4 + 4^4.$$

The smallest fifth power that is the sum of five *distinct* fifth powers is

$$72^5 = 19^5 + 43^5 + 46^5 + 47^5 + 67^5.$$

X. Gonze also discovered that

$$125^6 = 118^6 + 93^6 + 2 \times 78^6 + 48^6 + 42^6 + 18^6 + 2 \times 6^6.$$

These examples do *not* improve on the bounds for $s(4)$, $s(5)$, and $s(6)$, but they are remarkable for the small numbers involved.

FURTHER READING

V. A. Demjanenko, "L. Euler's Conjecture", *Acta arithmetica*, 25 (1973–4), pp. 127–35

Noam Elkies, "On $A^4 + B^4 + C^4 = D^4$", *Mathematics of Computation*, 51 (1988), pp. 825–35

L. J. Lander and T. R. Parkin, "Counterexamples to Euler's Conjecture on Sums of Like Powers", *Bulletin of the American Mathematical Society*, 72 (1966), p. 1079

L. J. Mordell, *Diophantine Equations* (New York: Academic Press, 1969)

Paulo Ribenboim, *13 Lectures on Fermat's Last Theorem* (New York: Springer-Verlag, 1979)

Ian Stewart, *The Problems of Mathematics* (Oxford: Oxford University Press, 1987)

David Wells, *The Penguin Dictionary of Curious and Interesting Numbers* (Harmondsworth: Penguin Books, 1986)

9

Pascal's Fractals

This is a True Life Story. No names have been changed to protect the innocent, since – as Kurt Vonnegut pointed out in *The Sirens of Titan* – the protection of the innocent is a matter of heavenly routine.

Inspiration often comes from unexpected sources. Some time ago I was reading Gregory Chaitin's *Algorithmic Information Theory*, a remarkable and stimulating book about randomness in the logical structure of arithmetic. I don't want to talk about that here; in fact, I want to talk about *regularities* in the structure of arithmetic. Anyway, Chaitin's book includes a diagram, which I recognized – and a theorem, which I didn't.

The diagram, and the theorem, are about Pascal's triangle. This is a triangular array of numbers (figure 9.1) whose left and right hand borders are all 1's, and where each number is the sum of the two immediately above it. Symbolically,

The kth number in the nth row (counting both n and k starting at zero) is the *binomial coefficient* $C(n,k)$. These numbers occur as the coefficients of x^k in the expansion of $(1 + x)^n$, hence the name. For example,

$$(1+x)^4 = 1 + 4x + 6x^2 + 4x^3 + x^4,$$

corresponding to row 4 of Pascal's triangle. Binomial coefficients are important throughout mathematics.

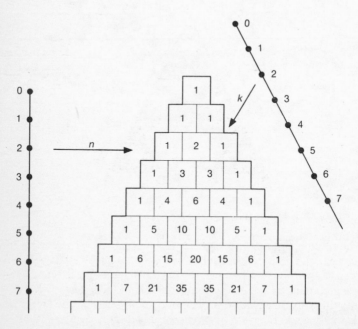

9.1   *Pascal's triangle: each number is the sum of the two above it.*

Sometimes binomial coefficients are even, and sometimes they are odd. How can you tell which? That's what the diagram and theorem in Chaitin's book were about.

Draw Pascal's triangle as a grid of squares, like bricks in a triangular wall. Colour a square black if the corresponding number is odd, and white if it is even. To generate such a picture, it is *not* necessary to work out the exact numbers in Pascal's triangle! All you need is the symbolic rule above, together with the information that

> 1 is odd,
> odd + odd = even + even = even,
> odd + even = even + odd = odd.

In other words, you colour the squares down the two sides black, and then colour squares white if the two above them have the same colour, black if they have different colours. It doesn't take long to fill in the entire triangle.

9.2   *Patterns in Pascal's triangle: white = even, black = odd.*

The result is figure 9.2, a dramatic and intricate pattern of black and white triangles. It closely resembles the *Sierpiński gasket* (figure 9.3). To make a Sierpiński gasket you start with a black triangle. Divide it into four equal triangles and colour the one in the middle white; then repeat on the three smaller black triangles, and so on forever. The Sierpiński gasket belongs to the class of geometrical objects known as *fractals*. A fractal is something that has detailed structure however much it is magnified. For example, if the surface of a perfect sphere is magnified, it

9.3   The Sierpiński gasket, a fractal derived from a triangle by repeatedly dividing it into four.

becomes flat and featureless: it is not a fractal. But the shrinking triangles of the Sierpiński gasket go on forever, and the object looks complicated however much it is magnified. So the Sierpiński gasket is a fractal.

If you draw a black-and-white Pascal's triangle with an enormous number of rows, and look at it from a large distance, then it looks just like a Sierpiński gasket. This fact has a curious consequence.

Among the integers, odd and even numbers occur equally often. A number picked at random will be even with probability $\frac{1}{2}$, and odd with probability $\frac{1}{2}$. You might expect the same to be true of the numbers in Pascal's triangle: half even, half odd. Now the probability of getting an even number in Pascal's triangle is the proportion that is coloured white in figure 9.2, and the probability of getting an odd number is the proportion coloured black.

For larger and larger numbers of rows in Pascal's triangle, these two probabilities are approximated better and better by the corresponding proportions of a Sierpiński gasket. So we may ask: *what proportion of a Sierpiński gasket is white?* Stop and think about it for a moment before I give the game away.

OK, here goes. Consider how the gasket is constructed. Start with a black triangle of total area 1. Paint an upside-down triangle, one quarter the size, white. This leaves three smaller black triangles, each of area $\frac{1}{4}$, and the remaining black area has shrunk from 1 to $\frac{3}{4}$. Now paint an upside-down white triangle on each of these: the black area shrinks to $\frac{3}{4} \times \frac{3}{4}$. Repeat indefinitely. More and more of the gasket gets painted white, and the black area becomes $\frac{3}{4} \times \frac{3}{4} \times \ldots \times \frac{3}{4}$, which tends to *zero* as the number of stages becomes very large.

In other words, the black part of a Sierpiński gasket has total area 0, the white part has area 1.

In terms of Pascal's triangle, this means that *almost all numbers in it are even*. In a very large Pascal's triangle, odd numbers occur with probability very close to zero. So we've learned something surprising about Pascal's triangle, by thinking fractally about the Sierpiński gasket.

What about generalizations? Odd and even are special cases of "arithmetic to a modulus". Let me remind you what that involves.

Choose a number, for example 5, and call it the *modulus*. Replace all numbers by their *remainders* on division by 5. Since remainders must be less than 5, only the numbers 0, 1, 2, 3, 4 survive. It is still possible to do arithmetic in this depleted number system. You can add any two such numbers, if you remember again to replace the sum by its remainder after dividing by 5. The result is called *arithmetic to the modulus 5*, or more briefly, (mod 5). In this brand of arithmetic, 2+2 = 4 as usual, but 3+4 = 2. That's because 3+4 = 7, and the remainder on dividing 7 by 5 is 2. The addition table (mod 5) is:

+	0	1	2	3	4
0	0	1	2	3	4
1	1	2	3	4	0
2	2	3	4	0	1
3	3	4	0	1	2
4	4	0	1	2	3

You can use any modulus, not just 5; and in fact you can also do multiplication, though we won't need this.

The distinction between odd and even numbers is just arithmetic (mod 2). An even number leaves remainder 0 on division by 2; an odd number leaves remainder 1. So all even numbers are 0 (mod 2) and all odd numbers are 1 (mod 2).

Thus we can generalize figure 9.2 by asking "What does Pascal's triangle look like (mod 5)?", or indeed the same question with "5" replaced by any other number. The results, for various moduli, are shown in figure 9.4. Here white squares correspond to numbers that are 0 to the chosen modulus (that is, exact multiples of the modulus), and all other values are black. You can generate these pictures yourself, using the symbolic rule for forming Pascal's triangle, but carrying out the

3

4

5

6

7

8

9

10

12

9.4 Patterns in Pascal's triangle to the moduli 3, 4, 5, 6, 7, 8, 9, 10, 12. The patterns for prime moduli (3, 5, 7) tend to be simpler.

addition using the table for arithmetic to your chosen modulus. Again you'll observe striking patterns of triangular regions.

You may enjoy producing your own patterns. If so, there are many ways in which you can explore new territory. You could:

1 *Change the colouring rule.*
   For example: what happens if you paint the kth cell in row n black when $C(n,k)$ is 1 (mod 5)? Or, more ambitiously, if you use a colouring scheme 0 = white, 1 = red, 2 = yellow, 3 = blue, 4 = black?

2   *Change the modulus.*
    What happens (mod 3)? (mod 11)? (mod 1001)?
3   *Change the rules.*
    Start with different numbers down the sides of the triangle, not just
    1's. Make each number the difference of the two above, not the sum,

Or make it the sum of the number on the left and twice that on the right:

You don't need a computer to do any of these. You can easily get thirty
or forty rows by hand as long as the modulus isn't too big.

Anyway, that deals with the diagram in Chaitin's book. But the
theorem was, if anything, even more intriguing. It lets you *predict*
whether a cell will be black or white, *without* calculating the corresponding binomial coefficient.

To explain the theorem I need another idea: the representation of
numbers to a given *base*. The usual way to write numbers is base 10 (or
decimal) notation. This means that, for instance,

$$321 = (3 \times 10 \times 10) + (2 \times 10) + (1 \times 1).$$

In base 7, say, the same symbol 321 would mean

$$(3 \times 7 \times 7) + (2 \times 7) + (1 \times 1),$$

which is 162 in decimal.

In particular, base 2 notation, or *binary*, is what computers use. In
binary the only digits are 0 and 1. Here's a short table of binary notation.

Decimal	*Binary*
0	0
1	1
2	10
3	11
4	100
5	101
6	110
7	111
8	1000
9	1001
10	1010

Suppose we have two numbers n and k, in binary notation. Write them one above the other with corresponding digits aligned. For instance if n = 1001 (9 decimal) and k = 101 (5 decimal) then write

    1001
     101.

Say that k *implies* n, written $k \to n$, if every binary digit of the bottom number k is less than or equal to the digit of n above it. In other words, $k \to n$ if you *never* get a pair of corresponding digits looking like

    0
    1.

Write $k \nrightarrow n$ otherwise. (The curious use of "implies" comes from computer logic operations.)

For example, to see whether $9 \to 5$, we look at what I wrote a few lines back, and see that the third column from the right has 0 on top, 1 below. So $9 \nrightarrow 5$. On the other hand $21 \to 23$ since in binary these are

    10111
    10001

and no digit on the bottom is greater than the digit above it.

The theorem stated by Chaitin – and, as he says, originally proved by the great French recreational mathematician Edouard Lucas a century ago – is as follows:

Lucas's theorem

$C(n,k)$, the kth entry in row n of Pascal's triangle, is:
   even if $k \nrightarrow n$,
   odd if $k \to n$.

It gives a quick and efficient way to test the parity (oddness or evenness) of $C(n,k)$. For example, since $21 \to 23$, it follows that $C(23,21)$ must be odd. In fact $C(23,21) = 253$. To find the parity of, say, $C(17,5)$, we express them in binary:

    17 = 10001
     5 =   101.

Pad out the bottom number with zeros to make them the same length:

    17 = 10001
     5 = 00101.

The third column from the right means that 5↛17, so C(17,5) is even. In fact C(17,5) = 24752.

You can check Lucas's theorem for other cases if you want: I'm not going to prove it here. It's a remarkable result, because it relates the arithmetic of Pascal's triangle to the *modulus* 2 and notation to *base* 2. Significant properties of numbers do not usually depend on their digits in some notational system – but here they do.

Does Lucas's theorem generalize to moduli other than 2? Let's try to find out.

The first interesting case is the modulus 3. The pattern of Pascal's triangle (mod 3) is shown in figure 9.5. Here the kth cell in row n is shown:

> white   if C(n,k) = 0 (mod 3),
> black   if C(n,k) = 1 (mod 3),
> with a dot if C(n,k) = 2 (mod 3).

Clearly there's some structure to the problem: the pattern is by no means random. But can we predict the entire pattern, rather than just calculate bits of it?

9.5   Pascal's triangle (mod 3). White = 0 (mod 3), black = 1 (mod 3), dot = 2 (mod 3).

Lucas's theorem relates C(n,k) (mod 2) to the digits of n and k (base 2). It seems natural to try to relate C(n,k) (mod 3) to the digits of n and k (base 3). Let's experiment. Try row n = 11 (decimal), which is 102 in base 3. The pattern of values is

k	0	1	2	3	4	5	6	7	8	9	10	11	(decimal)
$C(11,k)$	1	2	1	0	0	0	0	0	0	1	2	1	(mod 3).

Writing n and k to base 3, and collecting together cases that give the same answer, we find:

$C(n,k) = 0$ (mod 3):

n	102	102	102	102	102	102
k	010	011	012	020	021	022

$C(n,k) = 1$ (mod 3):

n	102	102	102	102
k	000	002	100	102

$C(n,k) = 2$ (mod 3):

n	102	102
k	001	101.

For convenience I've again padded out k with extra zeros on the left, so that every number has the same number of digits.

Exactly as in binary, let $k \rightarrow n$ mean "every digit of k is less than or equal to the corresponding digit of n"; but this time use digits to base 3. Here $k \rightarrow 102$ only when k is 000, 001, 002, 100, 101, and 102. Comparing with the above results, we find that these are precisely the cases when $C(n,k)$ is 1 or 2 (mod 3). In other words, $C(n,k)$ is 0 (mod 3) – that is, a multiple of 3 – if and only if $k \nrightarrow n$, when n and k are written to base 3.

We haven't *proved* this; but if you check it experimentally you'll find that it always works. It can be considered as generalizing Lucas's theorem, because that says that $C(n,k)$ is 0 (mod 2) if and only if $k \nrightarrow n$ (base 2). We've just changed "2" to "3" throughout.

However, if a number isn't even then it must be odd. So Lucas's theorem gives complete information (mod 2). Our generalization does *not* give complete information (mod 3), because when $k \rightarrow n$ the value might be 1 or 2 (mod 3). How can we decide which?

Before reading on, you may like to try your hand unaided. I'll give you a clue. Say that a pair of corresponding digits of n and k is *crucial* if n has digit 2 and k has digit 1. Take note of the number of crucial pairs.

Here's my answer: a first attempt at a Grand Lucas Theorem. It looks *very* strange!

Lucas's theorem (mod 3).

$C(n,k)$ (mod 3) is:
  0 if $k \not\to n$.
  1 if $k \to n$ and the number of crucial pairs of digits is even.
  2 if $k \to n$ and the number of crucial pairs of digits is odd.

For example, if you want to know $C(62,30)$ (mod 3) you just write

$n = 62$ decimal $= 2022$ base 3
$k = 30$ decimal $= 1010$ base 3.
          ↑ ↑

First, $k \to n$ so the result is 1 or 2. There are two crucial pairs, marked by arrows. Since two is even, we know that $C(62,30)$ must be 1 (mod 3).

This is striking. $C(62,30)$ is a number with eighteen digits, and I don't know what it is exactly. But I do know what it is (mod 3)!

For a more accessible check on the theorem, let's try $C(14,10)$. We have

$n = 14$ decimal $= 112$ base 3
$k = 10$ decimal $= 101$ base 3.
                ↑

Again $k \to n$, but there is only one crucial pair. Since one is odd, $C(14,10)$ must be 2 (mod 3). In fact $C(14,10) = 1001 = 999+2 = 3 \times 333 + 2$, so I'm right.

So at least we've got an *answer* for modulus 3, but it's all rather baffling. Why does it depend on crucial pairs? Mathematics is about understanding, not just answers: what's *really* going on?

When I started thinking about other moduli, I couldn't see how to make anything like crucial pairs work. I got stuck. This is not uncommon in mathematical research. One trick of the trade is to pick other people's brains. So I mentioned the (mod 3) result to a colleague, John Jones. He's a topologist, so I thought it was a fair bet that he hadn't come across this theorem in combinatorial number theory. But mathematics is full of surprises.

"Oh, yes," he said. "That sort of thing's very important in topology. The answer for any prime modulus is given in *Cohomology Operations* by Epstein and Steenrod." He was right. (David Epstein's office is just down the corridor from mine. Isn't life full of strange coincidences? Read on, there's more.) The next day Mike Paterson, from the Computer Science Department, told me that the result is in volume I of Donald Knuth's monumental classic *The Art of Computer Programming*. Two days later, volume 10 number 2 of the *Mathematical Intelligencer* magazine arrived – and there on p. 56 was a long article by Marta Sved, with much the same pictures as I've drawn here, the general theorem for any modulus, and information on related questions such as Stirling numbers.

All this drives home to me the unity and diversity of mathematics, the problems of the information explosion, and the inescapable perversity of the universe.

Be that as it may, let me tell you what the genuine Grand Lucas Theorem is. It tells you how to compute $C(n,k)$ to any *prime* modulus p. And the whole idea is far, far neater than crucial pairs, which suddenly look chewing-gum-and-stringy compared to the deeper pattern.

I'll describe the theorem by example. Suppose you want to find $C(216, 159)$ (mod 7). First, write 216 and 159 to base 7. That is, express them as $a \times 49 + b \times 7 + c$ and write them in the notation abc. The result is that 216 (base 10) is equal to 426 (base 7), and 159 (base 10) is equal to 315 (base 7). Write these one underneath the other,

426
315

and form the three binomial coefficients given by the columns: $C(4,3)$, $C(2,1)$, and $C(6,5)$. Work these out,

$C(4,3) = 4$
$C(2,1) = 2$
$C(6,5) = 6$

and multiply the results to get $4 \times 2 \times 6 = 48$. Finally reduce this (mod 7) to get 6. *This is the answer.*

It works for any prime modulus. If you get "impossible" binomial coefficients $C(n,k)$ where n is smaller than k, you must treat them as being zero.

How does this fit with the now discredited idea of crucial pairs, which solved the problem (mod 3)? The answer is that any crucial pair contributes a factor of 2 to the value we want, whereas non-crucial pairs always contribute a factor 1. (When $k \to n$ no zero factors occur.) Now $2 \times 2 = 1$ (mod 3). Pairs of 2's "cancel out". So the product of all the factors is 1 if there is an even number of 2's, and 2 otherwise. Crucial pairs are red herrings!

And what *that* shows is: don't always be satisfied with the first pattern you find. There may be a deeper and better explanation.

Here are some questions for you to think about.

1. Choose any modulus m. For numbers k and n written in base m notation, define $k \to n$ if every digit of k is less than or equal to the corresponding digit of n. One possible generalization of Lucas's theorem (mod 2) is: $C(n,k)$ is 0 (mod m) if and only if $k \not\to n$. Is this true?

2. If not, for which *n is* it true?
3. How can you determine C(*n,k*) (mod 4)?
4. How can you determine C(*n,k*) (mod 5)?
5. How can you determine C(*n,k*) (mod 6)?

ANSWERS

1   No. For instance, in base 4 we have 2 $\not\to$ 4 because 2 is 02 and 4 is 10 in base 4; but C(4,2) = 6 which is 2 (mod 4), not 0.

2   The Grand Lucas Theorem shows that the answer includes all *prime* moduli.

3   Because $4 = 2^2$, this comes under the general heading of prime power moduli. The results in this case are equally satisfying but much more complicated, and I refer you to Marta Sved's article, listed below.

4   This is a direct consequence of the Grand Lucas Theorem, because 5 is prime.

5   This can be answered by breaking it into two parts. You can always work out what a number is (mod 6) provided you know it (mod 2) and (mod 3). Here's a table that shows how:

mod 2	mod 3	mod 6
0	0	0
0	1	4
0	2	2
1	0	3
1	1	1
1	2	5

For instance, a number is 5 (mod 6) if and only if it is 1 (mod 2) and 2 (mod 3). So we can find C(*n,k*) (mod 6) by combining its values (mod 2) and (mod 3). By the Grand Lucas Theorem, these can be found by writing *n* and *k* to base 2 and then to base 3.

So now two different number bases have come into the picture! I'd be surprised if you can answer question 5 using only the expansions of *n* and *k* to base 6 – let me know if you manage it.

FURTHER READING

Gregory J. Chaitin, *Algorithmic Information Theory* (Cambridge: Cambridge University Press, 1987)

Martin Gardner, *Mathematical Carnival* (Harmondsworth: Penguin, 1978)

Benoît Mandelbrot, *The Fractal Geometry of Nature* (San Francisco: Freeman, 1982)

Alfréd Rényi, *A Diary in Information Theory* (Cambridge: Cambridge University Press, 1987)

Ian Stewart, *Concepts of Modern Mathematics* (Harmondsworth: Penguin, 1981)

Marta Sved, "Divisibility – with Visibility", *Mathematical Intelligencer*, 10/2 (spring 1988), pp. 56–64

10

The Worm Returns

Henry Worm was coiled up in his favourite armchair by the fireside, reading the financial pages of the newspaper. "Woodworm's shares are down a point today, Anne-Lida . . . Maggots and Spencer are doing better, though, and Slime Darby Holediggings are really going great . . ."

"Henry, you know very well that we don't own any shares! Now put that paper down, and apply your mind to something *important*."

"Yes, dear," said Henry, thinking that *the early hen pecks the worm* as he folded the newspaper sadly. "What important question did you have in mind?"

"The question of my sister Worma's birthday-present," said Anne-Lida.

"Ah. Yes. How about a pair of socks?"

"What would a worm want a pair of socks for?"

"A sock, then. A knitted bedsock."

"We gave her a bedsock last year, Henry."

"The left or the right?"

"The left."

"Then this year we can give her the right sock."

"Henry, we gave her the right sock two years ago. No, I want to send her something more thoughtful, something we have taken more *trouble* over."

"Something *I* have taken more trouble over," muttered Henry. "Nothing, dear. Don't worry, leave it to me. I'll think of something really unusual!"

"That's what worries me," said Anne-Lida.

A week later, Henry staggered into the hole carrying a huge parcel. Gift-wrapped.

"What under the earth is *that*?"

"Worma's present, my sweet."

Anne-Lida peered at the object disdainfully. "Well, at least you got something *big*, whatever it is," she said. "Not like the earrings you bought in 1962, with the broken clasps."

"I didn't think the clasps would matter. After all, worms don't have ears. You must admit they fitted her beautifully!"

"Yes. Like a pair of lifebelts. But let us not argue, Henry. What did you buy?"

Proudly Henry unwrapped the present.

Anne-Lida's tail drooped. "Pizza?"

"The biggest pizza in the world, my pet. Fit for a queen!"

"The crust seems rather *thin*."

"Thinnest in the world. Thinner than paper. Exquisite when baked!"

"Hmmph. But Henry, it must be a metre across, at the very least."

"*Exactly* a metre, Anne-Lida."

Anne-Lida sniffed. "Well, don't blame me if you have trouble posting it."

"Sending pizza through the post," said Henry, "is no problem . . ."

"That will be £5," said Hector the postworm. He paused, and eyed the parcel warily. "Unless . . . H'excuse me, sir, but how big is that parcel?"

"One metre in width."

"What do you mean, width?"

"The maximum distance in a straight line between any two of its points."

"Oh dear."

"What do you mean, 'Oh dear'?"

"I think you h'are referrin' to what we in the Post Office technically call the *diameter* (figure 10.1). Now I am h'obliged to inform you, sir, that if the diameter was *less* than 1 metre, there would be no problem. H'unfortunately the postal rules state that parcels will be accepted only if their diameter is *less* than 1 metre." He took out a gigantic pair of callipers and measured the parcel carefully. "Yes, just as I feared. One metre *precisely*. Not less, sir: equal. I'm afraid I'm not permitted to h'accept it."

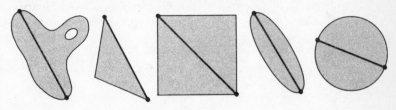

10.1   The diameter *of a plane figure (or one in higher dimensions) is the largest distance in a straight line between any two of its points.*

"Look, let's be reasonable," said Henry. "I can always shave a little bit off."

"Yes, sir, but that may not reduce the maximum width. For h'instance, sir: suppose the pizza was a perfect circle, 1 metre in diameter. Then shaving a little bit off would make it less than a metre across in one direction – but in others it would still be a metre across." (Figure 10.2)

"I don't think it's *exactly* circular," said Henry. "It looks a bit irregular to me."

"Same principle," said Hector airily. "Of course, you could always cut it up and send it in pieces."

"Yugh," said Henry. "Sorry. Talk of cutting things up always makes me feel queasy. My mother had a near miss with a lawnmower and I was nearly born as twins . . . But you're right, I could – ugh – slice it in half . . ."

"That might not work, sir. If you slice a metre circle in half then at least one piece will still have diameter 1 metre." (Figure 10.3)

"Well, quarters then. That would do it (figure 10.4). I don't know offhand what the diameter is, but it's surely less than 1 metre."

"Excellent. That will be £20, sir."

"What!?" screeched Henry Worm in dismay.

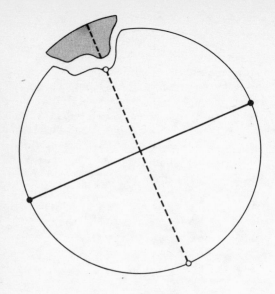

10.2 *Cutting off a small piece of a circle does not decrease its diameter.*

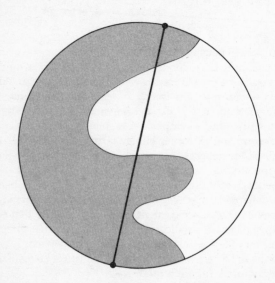

10.3 *If a unit circle is cut into two pieces, at least one piece contains two diametrically opposite points and hence has diameter 1.*

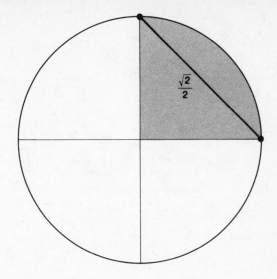

10.4 *A circle of unit diameter cut in four yields pieces with diameter* $\frac{\sqrt{2}}{2} = 0.7071\ldots$, *smaller than 1 . . .*

"Flat rate, sir: £5 per package."

"I'm not spending £20 to send my sister-in-law a pizza! It only cost £3!"

"Then you'll have to cut it into fewer pieces. If I were you, I'd cut it up into the smallest number possible."

"Good idea!" said Henry. *The first helpful thing the idiot's said.* "What number would that be?"

The postworm stared at the ceiling. "Three, for a circular pizza, sir (figure 10.5). But in general I haven't the foggiest h'idea. Depends what *shape* it is, you see."

"Don't worry about the shape. What's the smallest number that will work for *any* shape? I can see how to do it in *four* pieces, but maybe that's too big. On the other hand, your circle example shows that *two* pieces aren't enough . . . So the smallest number has to be either three or four."

"I still haven't the foggiest h'idea," said the postworm.

"Neither have I," said Henry. "But, unlike you, I have an enquiring and intelligent mind. I shall find out . . ."

Henry Worm sat coiled in his favourite chair, making irritated annotations on a piece of paper.

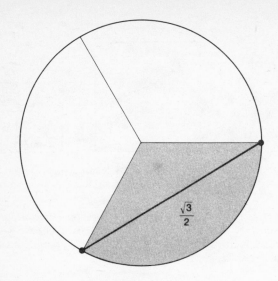

10.5   ... *but a circle of unit diameter cut in three yields pieces with diameter* $\frac{\sqrt{3}}{2} = 0.8660...$, *which is also smaller than* 1.

"Henry dear, *what* under the earth are you scribbling for?"

"Er – it's a mathematical problem, my dear."

"Oh."

"Given a shape in the plane, of unit diameter, you must cut it up into several pieces, each of diameter *less* than 1 unit."

"Dice it, Henry."

"Yes dear, that would work, but it would cost a fortune in postage!"

"Henry, what are you . . ."

"I want to find the smallest number of pieces that will always work, no matter what the shape may be. *That*, my dear, is what I was 'scribbling' about. A question of the highest intellectual depth."

"Henry, you're talking nonsense."

"Yes, dear. The answer is either three or four."

"How under the earth can you tell that, Henry? You said yourself you don't know what the shape is!"

"Ah, but I can prove that every figure of diameter 1 metre is contained inside a square of side 1 metre. To see why, fit the shape into a right-angled corner (figure 10.6(a)) as tightly as possible. Then it can't project

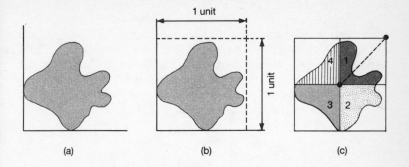

10.6 *(a) Any shape fits snugly into a right-angled corner. (b) If it has diameter 1 then it cannot cross two lines parallel to the walls, one unit distant. Therefore it lies inside a unit square. (c) Such a square can be cut into four pieces, each of diameter $\frac{\sqrt{2}}{2} = 0.7071\ldots$; hence so can the original shape.*

over two lines drawn parallel to the walls, but 1 metre away (figure 10.6(b)). Because, if it did, it would have diameter *more* than 1 metre."

"Agreed. That's obvious, Henry."

"When ideas of genius are pointed out with sufficient clarity, even the feeblest wit can comprehend them," said Henry.

"That's as may be. But your genius-level brain hasn't yet explained with sufficient clarity why you wish to surround the shape with a square."

"Ah . . . Well, if I cut the metre square into four parts, then the shape is also cut into four parts (figure 10.6(c)). Each has diameter at most $\frac{\sqrt{2}}{2} = 0.7071\ldots$ metres, which is less than 1. Because that's the length of the diagonal of the smaller squares."

"You know, Henry, there are times when you're almost as clever as you think you are . . . Can you cut a square of side 1 metre into *three* pieces of smaller diameter than 1?"

"I don't *think* so," said Henry. (*He's right. Can you prove it?*)

"Perhaps you could replace the square by something smaller, which *would* cut into three pieces of diameter less than 1 metre. Something that doesn't stick out so much at the corners."

Which, Henry Worm was forced to admit, wasn't a bad idea at all . . . But what shape should he use? He returned to his pad and began doodling. Soon an idea began to crystallize.

"Anne-Lida, I do believe a regular hexagon will work! Look, suppose I can surround the shape with a hexagon whose sides are 1 metre apart (figure 10.7(a)). Then I can cut it into three parts whose diameters are less than 1 metre (figure 10.7(b))." (*What is the diameter of the three pieces?*) "I'm not sure if such a hexagon exists in general, but . . ." (tries it) " . . . it certainly does for Worma's pizza, look!"

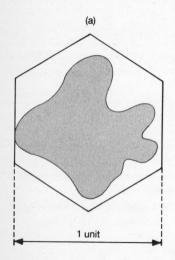

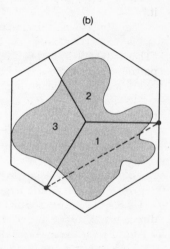

10.7   (a) Surround a shape of unit diameter by a regular hexagon whose opposite sides are 1 unit apart. (b) Cut into three pieces, each of diameter less than 1.

"You should post Worma's present now, Henry. Even if the general proof is lacking."

"Certainly, my dear."

"And Henry?"

"Yes, my little caterpillar?"

"Don't forget that it's mother's birthday next week. Try to think of something we can send her."

"A handkerchief?"

"Don't be silly, Henry. Worms have no noses."

I was thinking more of using it as a gag, thought Henry . . .

Two days later Henry staggered into the hole with a large gift-wrapped parcel.

"What under the earth is *that?*"

"Your mother's present, my sweet."

Anne-Lida peered at the object disdainfully. "Well, at least you got something *big*, whatever it is," she said. "What *is* it, by the way?"

Proudly Henry unwrapped the present.

Anne-Lida's tail drooped. "Cheese?"

"The biggest cheese in the world, my sweet."

"Henry, it must be a metre across, at the very least."

"Oh, surely not, Anne-Lida."

Anne-Lida sniffed. "Well, don't blame me if you have trouble posting it."

"Sending cheese through the post," said Henry, "is no problem . . ."

"That will be £5," said Hector the postworm. He paused, and eyed the parcel warily. "Unless . . . H'excuse me, sir, but how big is that parcel?"

"What do you mean, 'big'?"

"Its diameter, sir. The largest distance in a straight line between any two of its points, just the same as for two dimensions. Sir."

"Just under a metre," said Henry cheerfully.

"Let me just measure it, sir . . . Hmmm . . . I make it *exactly* 1 metre."

"Oh."

"Which poses what we in the Post Office call a bit of a problem, sir."

"Don't tell me. All right, cut it into three bits, then!"

Hector the postworm picked the parcel up and eyed it from several directions. "I'm not sure *three* pieces will do it, sir."

"But I've just proved that three . . ."

"Yes, sir. For plane figures, sir."

"Oh."

"This here package is what we in the Post Office refer to h'as a *bulky item*, sir. A three-dimensional parcel, if you catch my meaning."

"I'm beginning to see the problem, yes."

"Now, were this a spherical cheese, like a Gouda, shall we say, then you couldn't possibly comply with Post Office regulations by cutting it into three pieces." (*Why not?*) "On the other hand, four would certainly be enough." (Figure 10.8)

"Fine, quarter it, then."

"You haven't looked very closely at the figure, sir. Cutting it into quarters doesn't work."

"Well, eighths, then. That will do the trick! Every solid of diameter one metre fits inside a unit cube – the proof is much the same as for two dimensions – so if we cut that into eight equal cubes half the size, then the

10.8   How to cut a sphere of unit diameter into four pieces of diameter less than 1.

small cubes will have diameter less than 1 . . ." (*What is their diameter?*)
" . . . and so will the pieces of the original solid."

"Excellent, sir. That will be £40."

"*What?* But the cheese only cost £7! Look, I'll take it away and experiment a bit first . . ."

But the intricacies of three-dimensional geometry were too much for Henry. So he went to see a friend of his, an obscure clerk at the Patent Office named Albert Wormstein who seemed to have a bit of a mathematical mind. He found him standing in front of a blackboard, writing formulas on it and then rubbing them off again in irritation. "$E = ma^2$ . . . No, ridiculous! $E = mb^2$ . . .? Better, but not right, not right at all. $E = m$ . . . "

"Hello, Albert! Sorry, am I interrupting anything important?"

"Good heavens no, Henry. Just a little idea that won't quite work out . . . Nice to see you again." And Henry explained his problem.

"Borsuk!" said Albert.

"Same to you!" replied Henry, with some heat. So Albert had to explain that he'd meant no offence. Henry's problem was a question in combinatorial geometry: the arrangement of shapes. It was first posed by the Polish mathematician K. Borsuk in 1933, and is therefore known as the *Borsuk Problem*. It had been solved, Albert told him, for two- and three-dimensional bodies, but is still open for bodies of dimension four or more.

The Borsuk Problem asks for the smallest number of pieces such that any set of unit diameter in n-dimensional space can be cut into that many sets of strictly smaller diameter. In 1933 Borsuk proved that for figures in the plane, three pieces are always sufficient. His proof was the same as Henry's. Henry's conjecture, that every plane set of diameter 1 can be surrounded by a regular hexagon whose opposite sides are distance 1 apart, was proved by the Hungarian mathematician J. Pál in 1920. As Henry had realized, the solution to Borsuk's Problem in the plane then follows. For sets in the plane the answer to the Borsuk Problem is that three pieces are always enough.

Is the same true for sets in three-dimensional space? Borsuk made the same observation as Hector the postworm: the answer is "no". A sphere of diameter 1 cannot be divided into three pieces of diameter less than 1. In 1933 he conjectured that four pieces suffice for any set in three-dimensional space, and more generally that $n+1$ pieces suffice in n-dimensional space. But he couldn't prove his conjectures.

The first progress was made in 1955 by H. G. Eggleston, who proved that Borsuk was right in three dimensions. His proof, very long and difficult, was simplified in 1957 by Branko Grünbaum, using a similar trick to the Pál hexagon. In 1953 David Gale had proved a three-dimensional analogue of Pál's Theorem: every solid of diameter 1 can be surrounded by an octahedron in which opposite faces are distance 1 apart. Instead of a Pál hexagon, Grünbaum used a Gale octahedron. He showed that if three corners of the octahedron are cut off as in figure 10.9, then it can still contain any body of diameter 1. Finally he found a way to cut the resulting polyhedron into four pieces, each of diameter less than 1 (figure 10.10). The enclosed solid must also split up into four (or fewer) pieces, each of diameter less than 1.

"And what is the diameter?"

"The widest piece has diameter

$$\frac{\sqrt{(6129030 - 937419\sqrt{3})}}{(1518\sqrt{2})},$$

which is about 0.9887."

"Tricky," said Henry, impressed.

"Yes, combinatorial geometry is deceptive. There are lots of problems that look easy, but are still wide open. In a space of dimension four or higher, the Borsuk Problem still remains unsolved."

"Really?"

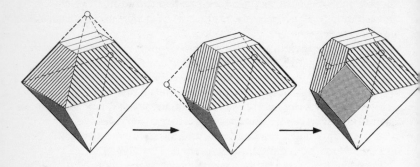

10.9   *Every solid of unit diameter can be surrounded by a* Grünbaum polyhedron, *obtained by slicing pieces off a* Gale octahedron *(one whose opposite faces are 1 unit apart).*

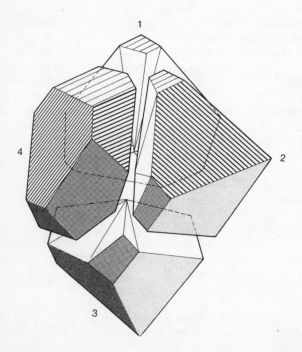

10.10   *The Grünbaum polyhedron can be cut into four pieces, each of diameter less than 1.*

"Indeed. It's known that *at least* $n+1$ pieces are required in general, but it's not known whether more than $n+1$ might be needed to cut some carefully chosen set of diameter 1 into pieces of smaller diameter. It *is* known that such a set must have sharp corners: in 1946 H. Hadwiger proved Borsuk's Conjecture for convex sets with smooth boundaries."

"So even in four dimensions, Borsuk's Conjecture is unproved? Let me see, that would say that any set in 4-space of diameter 1 can be cut into five pieces, each of diameter less than 1. Hmm . . . Almost sounds worth having a go at that! I bet I could get fairly close . . ."

"Possibly," said Albert.

"Can't you try the same argument I used with the square and the cube? Surround the shape by a unit hypercube and cut that into sixteen pieces half the size?"

"No, Henry, that method doesn't work any more. It's not easy, thinking in four-dimensional space." (*Why doesn't the method work? Use the fact that the "hyperdiagonal" of a four-dimensional "hypercuboid" of sides* a, b, c, d *is given by* $\sqrt{(a^2 + b^2 + c^2 + d^2)}$, *generalizing the Pythagorean theorem*.) "By cutting a unit hypercube into thirds you *can* show that eighty-one pieces are enough. But that's much too big."

"Are there any general results that work for all dimensions?" asked Henry.

"Indeed," said Albert Wormstein, nodding vigorously. "L. Danzer proved that in n dimensions

$$\sqrt{\frac{(n+2)^3}{3}} \cdot \left(2 + \sqrt{2}\right)^{\frac{(n-1)}{2}}$$

pieces are required. That comes to fifty-five when $n = 4$. But I'm sure you can do better than *that* in the four-dimensional case." (*Can you?*)

"Let me think about it . . . Anyway, Albert, thank you for your help."

"It was nothing. Now, where was I?"

"Er . . . $E = md^2$, I think."

"Thanks. Well, anyone can see *that's* no good! So next comes $E = me^2$ . . . No, no, terrible, terrible. You know, just before you arrived I really thought I was getting close, but now it feels as if I'm missing something important . . ."

ANSWERS

1. You can't cut a unit square into three pieces of diameter less than 1, because at least two of the four corners must belong to the same piece, and those are 1 or more units apart.

2. The three pieces of the hexagon have diameter $\frac{\sqrt{3}}{2} = 0.8660\ldots$.

3.   If you cut a sphere into three pieces then at least one of them must contain two diametrically opposite points. The proof of this is quite long, but requires no technical knowledge. See V. Boltjansky and I. Gohberg, *Results and Problems in Combinatorial Geometry*, p. 10.

4.   Each half-sized cube has diameter $\sqrt{((\frac{1}{2})^2 + (\frac{1}{2})^2 + (\frac{1}{2})^2)} = \sqrt{(\frac{3}{4})} = \frac{\sqrt{3}}{2} = 0.8660\ldots$.

5.   Henry's cube-halving method fails in four dimensions, because the diagonal of a half-size hypercube is $\sqrt{((\frac{1}{2})^2 + (\frac{1}{2})^2 + (\frac{1}{2})^2 + (\frac{1}{2})^2)} = 1$.

6.   Cut the hypercube in half in three directions but into three equal parts along the fourth. This yields twenty-four smaller hypercuboids, each of diameter $\sqrt{((\frac{1}{2})^2 + (\frac{1}{2})^2 + (\frac{1}{2})^2 + (\frac{1}{3})^2)} = \sqrt{(\frac{31}{36})} = 0.9279\ldots$. But perhaps you can improve on *that*?

FURTHER READING

V. Boltjansky and I. Gohberg, *Results and Problems in Combinatorial Geometry* (Cambridge: Cambridge University Press, 1985)

K. Borsuk, "Drei Sätze über die n-dimensionale Sphäre", *Fundamenta mathematicae*, 20 (1933), pp. 177–90

H. G. Eggleston, *Convexity* (Cambridge: Cambridge University Press, 1955)

Branko Grünbaum, "A Simple Proof of Borsuk's Conjecture in Three Dimensions", *Proceedings of the Cambridge Philosophical Society*, 53 (1957), pp. 776–8.

I. Yaglom and V. Boltjansky, *Convex Sets* (New York: Holt, Rinehart and Winston, 1961)

11

All Parallels Lead to Rome

The city that invented apartment blocks has an insoluble housing shortage.

The city that invented the public sewer has no adequate sewerage system.

The city which in 45 BC banned wagons from its centre during daylight hours has an average traffic speed of 6 km/hr. There are three cars for every metre of road.

The city is noisy, filthy, and heavily in debt – and one of the most beautiful in the world. A living paradox. No wonder that a one-sided surface is named after it.

No, there's no city named "Möbius". This is *Rome*. And when in Rome . . .

We sat at a table in the Via Vittorio Veneto, which winds downhill from the gardens of the Villa Borghese until it runs into the Piazza Barberini. An empty chianti bottle lay among the remains of a pasta lunch. A second, half full, stood beside it.

"Good job chianti doesn't come in Klein bottles," I said.

"I know that *klein* is German for 'small'," said Enrico, "and I agree that it's a good job chianti doesn't come in small bottles – but why have you lapsed into German all of a . . .?"

"No," I said. "I didn't mean that. It was a mathematician's joke. A Klein bottle is one whose inside and outside are the same."

"It would save on corks," said Elena.

"No, it would always leak," said Enrico. Enrico and Elena Macaroni: Henry and Helen. But it sounds so much more elegant in Italian. He runs an art gallery, and she runs him.

"How can a bottle have its inside the same as its outside?" Elena asked, serious now.

"It's a complicated story," I said. "The truth is that it doesn't really *have* an inside or an outside . . . And it isn't really a bottle."

"That explains a great deal."

"Who was Klein?" asked Enrico.

"Felix Klein was one of the greatest mathematicians that Germany ever produced," I said. "He was the second person to invent a surface with only one side. The first was August Möbius." I took a paper napkin, tore off a narrow strip, and joined its ends with a half-twist. "See: a Möbius band (figure11.1). But the Möbius band has an edge. Klein's bottle, invented in 1882, has no edges, it's a closed surface." (Figure 11.2(a))

"And it has only one side?"

"Imagine trying to paint the surface. You start on what looks like the outside, and carry on painting the tube. But it bends round, passes through itself, and then kind of turns inside-out. At that point you find that you're painting what you originally thought was the inside. There's only one side to the surface: it all joins up."

"But that's because it passes through itself," said Elena.

"No, it's because it turns inside-out and then joins up. I admit that it has to pass through itself if you want to make a model in three-dimensional space. In four-dimensional space it *doesn't* cross itself, but it

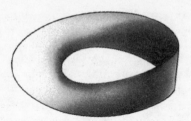

11.1 *The Möbius band.*

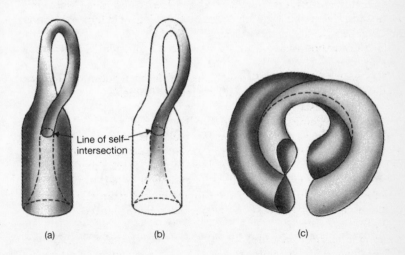

Line of self–
intersection

(a)   (b)   (c)

11.2 *Three views of the Klein bottle. (a) Embedded in 3-space. (b) Embedded in 4-space, the fourth dimension being illustrated by the depth of shading. The self-intersection in 3-space does not occur in 4-space (despite the way the picture looks when projected into 3-space as here): the positions in the fourth dimension of the two sheets of surface (that is, their shades) are different at the apparent intersection. (c) A less familiar form of the Klein bottle obtained by joining a figure 8 to itself with a half-twist. The shades distinguish the two lobes of the figure 8.*

still has only one side. Of course you have to learn how to think in four dimensions to see that." (Figure 11.2(b))

"Oh."

"Another way to obtain a Klein bottle is to take a figure 8, move it round a circle, and give it a half-twist as you do so (figure 11.2(c)). But that doesn't look very bottle-shaped. Actually," I went on, " I have a private theory about the name *Klein bottle*. I think it was originally *Klein's surface*. You see, in Klein's day there was quite an industry involving German mathematicians inventing new surfaces and getting them named after themselves. Kummer's surface and Steiner's surface, for instance, originally *Kummersche Flache* and *Steinersche Flache*, the '-sche' being a possessive ending and 'Flache' being German for 'surface'. So it probably started out as *Kleinsche Flache*, 'Klein's surface'. But it *looks* like a bottle, and the German for bottle is *Flasche*, so . . . "

"Some graduate student called it the *Kleinsche Flasche!*" said Elena. "'Klein's bottle'! A German pun!"

"Exactly. Or maybe it was mistranslated. I do know that in Hilbert and Cohn-Vossen's famous book *Anschauliche Geometrie* they refer to 'Klein's surface, also known as the Klein bottle'. Maybe Hilbert invented the pun."

"Fascinating," said Enrico. "Not very relevant to the real world, though."

"Don't be so sure," I said. "You're an art-dealer, right?"

"You know that."

"Italy is famous for beautiful paintings. Masaccio, Canaletto, Gozzoli, Veneziano, della Francesca. Wonderful perspective, right?"

"Perspective was invented in Italy."

"Perspective *drawing* was invented in Italy. The basic idea was discovered by Brunelleschi, in about 1420. And the geometry of perspective was published by another Italian, Alberti, in 1436, in his book *Della pittura*. It's called *projective geometry*, and it describes the way in which the eye sees the world. The basic surface is known as the *projective plane*. In the projective plane there are no parallel lines: any two lines meet at a single point." (Figure 11.3)

"Crazy."

"Furthermore, as Klein showed in 1874, the projective plane has only one side."

"Not so good for painting, then," said Enrico.

"No, you're wrong: you can get twice the size of painting on the same size of canvas!" Elena pointed out.

Curiously, the projective plane was invented long before the Klein bottle; but it's virtually unknown outside mathematical circles, whereas the Klein bottle is famous. Below, we'll examine some of the possible

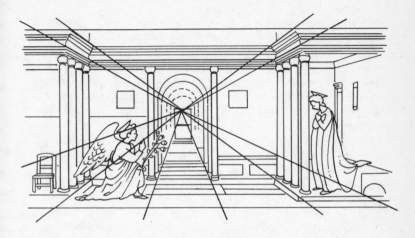

11.3 *The Annunciation of Domenico Veneziano (fl. 1438–61). The edges of the walls, in reality parallel, appear to the eye to meet "at infinity".*

reasons for this; but first we need to become familiar with the projective plane.

In ordinary geometry, there is a unique line joining any two distinct points. *Most* pairs of lines intersect in a unique point, but some – parallel lines – do not. But, from the right viewpoint . . .

I led Enrico and Elena from the Via Vittorio Veneto to the nearby Via XX Settembre, part of a long, straight stretch running for almost 4 km from the middle of Rome towards the suburbs.

"What do you see?" I asked them.

"Traffic. Jammed solid, as usual."

"No, I mean, something geometric."

"Nothing special."

"The two edges of the road, they're a pair of parallel lines. Parallel lines don't meet. Look at them: do they *look* as if they don't meet?"

Enrico and Elena humoured me by staring down the long, straight road.

"They do *seem* to meet," said Elena.

"On the horizon," said Enrico.

"Precisely," I said. "When the eye looks at parallel lines, they appear to meet. In the geometry of the visual system, parallel lines do not exist. So we need a new kind of geometry, in which *any* two lines meet.

"How far away is the point on the horizon where the two sides of the road would meet – if they were extended far enough?"

"Ooh, about 50 kilometres," said Elena.

"On a spherical Earth, yes. But on a plane?"

"Well . . . At the edge."

"It's a long way to the edge of a plane," said Enrico.

"Infinitely long," I said. "The place where parallel lines appear to meet is at infinity. In the usual Euclidean plane, infinity doesn't exist. You can go as far as you like, but you can't actually *get* to infinity. But in projective geometry, you can. To achieve this you have to add extra 'ideal' points 'at infinity' to the plane (figure 11.4). The points 'at infinity' form an extra line, so you have to add that too. What you get then is a slightly larger plane, so to speak, in which any two points are joined by a unique line and any two lines meet in a unique point."

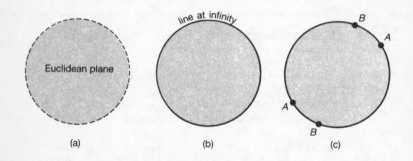

11.4 *The Euclidean plane (a) plus a line at infinity (b) forms the projective plane, provided we agree (c) that opposite pairs of points on the boundary such as* AA *or* BB *represent the same point of the projective plane.*

"But parallel lines meet in two points," said Elena. "One at one end, one at the other."

"Mmmm," I said. "But it would be nice if they met in only one, right? Prettier. More symmetric and elegant. More like actual lines."

"Yes," she said doubtfully.

"So we have to pretend that the two points at opposite ends of a pair of parallel lines are the same," I said.

"That's silly."

"Not as silly as it sounds. Have you ever been to infinity to see for yourself?"

"No."

"Mathematically, infinity is just an abstract construct, so we can endow it with any properties we want. I happen to want lines to meet in only one point. So I insist that the 'two' points at infinity, at either end of a pair of parallels, are to be considered as the same. It may sound odd, but it works. It's sort of like bending the lines round into a circle – except that they stay straight."

"Clear as mud."

"Good. So we get our first model of the projective plane: it's the usual plane, *plus* a 'line' at infinity, *plus* the rule that the opposite ends of pairs of parallels meet the 'line' at infinity in the same point." (Figure 11.5)

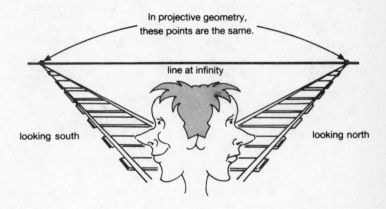

11.5 Looking south along a straight railway line we see two parallels meeting at infinity. Looking north, we see them meet again. Because two lines should meet in a unique point, we must identify these two "opposite" points at infinity.

"I'm having trouble visualizing it."

"On the contrary, Elena, it's how your visual system actually works."

"Well, I'm having trouble getting it into my head in one piece. And it's not at all clear to me why the projective plane has only one side, as you say. The ordinary plane has two sides: top and bottom."

"Yes, but the top surface and the bottom surface get joined together at infinity because of the rule about the end-points of parallels being the same," I said.

There are several different ways to "see" the shape of the projective plane, and some of them make it clearer than others that it has only one side. Probably the simplest is to take a topologist's viewpoint. As far as a topologist is concerned, the whole infinite plane can be squashed up inside a circular disc (figure 11.4(a)) – minus its boundary, of course. Then the extra "points at infinity" can be added in by gluing on the boundary as well (figure 11.4(b)). It looks circular, but that's not a problem to a topologist. To accommodate the rule about the opposite ends being the same, we have to (mentally) "glue" opposite points of the boundary circle together (figure 11.4(c)). If you try to bend the disc in three-dimensional space, so that this happens, then you have to pass it through itself (figure 11.6). The top half of the picture is called a *cross-cap*.

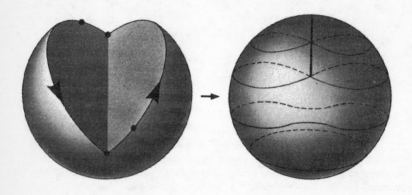

11.6   *If we attempt to identify opposite points in Figure 11.4(c) by physically bending the plane it is necessary for the resulting surface to pass through itself, forming a* cross-cap. *Along the self-intersection, the two "sides" of the plane join together to create a single-sided surface. The point at the top is* singular: *the surface near it cannot be continuously deformed into one or more separate discs.*

The cross-cap cuts through itself along a line. Just as for the Klein bottle, this line of self-intersection is an artefact caused by the way we draw the surface in three-dimensional space. Mathematically, it isn't "really" there. But it helps us to visualize it. To get rid of it, we should think of a disc, whose opposite boundary points are identified *mentally*, rather than by actually bending the disc around to bring them together.

You can see that this version of the projective plane only has one side. If you start painting the "outside" and cross the line of self-intersection, you end up on the "inside". You can see it in another way. If we cut out a strip that crosses the disc (figure 11.7(a)) then we really *can* glue the ends together – and we get a Möbius band (figure 11.7(b)). So the inside and outside already join up in this *part* of the projective plane. In fact we can

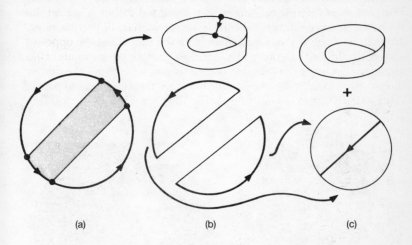

11.7 *A strip (shaded) that runs across the projective plane (a) forms a Möbius band (b) because opposite points on the boundary are identified. Suitably deformed, the remaining two pieces join to form a disc (c). Abstractly, we can form a projective plane by sewing a Möbius band and a disc together along their edges.*

see that a projective plane is just a Möbius band with a disc sewn on along the edge (figure 11.7(c)).

"It's all a bit unnatural," said Enrico.

"I agree," I told him. "Your sense of artistic elegance is functioning very well. But there's another model for the projective plane that is both geometric and natural. Of course, it has its own peculiarities."

"Of course."

"The idea is to increase the dimension of everything by one. When I say 'point' you must think 'line through the origin'. In ordinary three-dimensional space. When I say 'line' you must think 'plane through the origin'. When I say 'two points lie on a line' you must think 'two lines lie on a plane'. OK?"

"If it keeps you amused."

"Well, the abstract essence of geometry is the way that 'lines' and 'points' relate to each other, it's not their actual shape. The names are just useful labels. Here, it's useful to modify the labels to make the relationship clearer. In this 'beefed-up' version of geometry, any two 'points' lie on a unique 'line'. That is, any two lines through the origin lie in a unique plane (figure 11.8(a)). You agree?"

"Of course."

"But in addition, any two 'lines' meet in a unique 'point'. That is, any two planes through the origin meet in a line (figure 11.8(b)). So we have exactly the properties required in projective geometry. The projective plane is just three-dimensional space, but with a new meaning attached to 'point' and 'line'. Geometric, and natural."

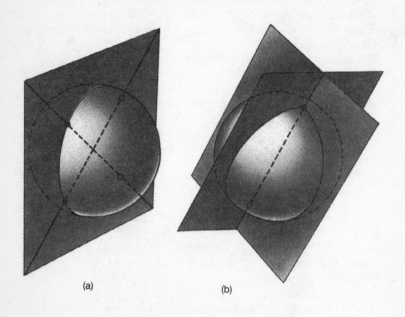

(a)   (b)

11.8   In ordinary 3-space, two lines through the origin determine a unique plane (a) and two planes through the origin determine a unique line (b). Each line cuts the sphere (shaded) in a pair of opposite points; each plane cuts it in a great circle. The projective plane can thus be interpreted either as the geometry of lines and planes through the origin in 3-space, or as that of point-pairs and great circles on a sphere.

"Natural?"

"Natural enough for a mathematician."

"But how can you call three-dimensional space a plane?" asked Elena.

"Because we've increased all the dimensions by one." I reminded her. "If a 'line' is a plane through the origin, then a 'plane' has to correspond to a three-dimensional object – which must be the whole space."

"Not only that: you can show that this new version of the projective plane is just the original one in disguise."

"How? It doesn't look like it to me!"

"It's a rather heavy disguise. Imagine a sphere centred at the origin. It cuts every 'point' of the projective plane – that is, every line through its centre – in a pair of opposite points. It cuts every 'line' – plane through its centre – in a great circle. So the geometry of the projective plane is just the geometry of the sphere, with 'point' interpreted as a pair of antipodal points, and 'line' interpreted as great circle."

"Fine. But we've got pairs of points, not individual ones."

"That doesn't really matter," I said. "Not in the abstract. But we can overcome that by thinking just of a hemisphere. That cuts most pairs down to single points."

"Except points on the boundary of the hemisphere."

"Precisely, Enrico! Well done! So we have to *identify* opposite points on the boundary of the hemisphere (figure 11.9). Just as our first model of the projective plane identified opposite points on the boundary of a disc.

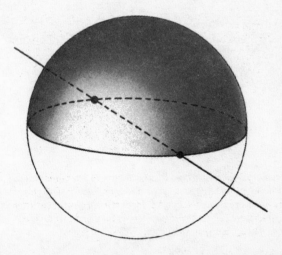

11.9 *To obtain single points rather than point pairs we can restrict attention to a hemisphere, obtaining a geometry of points and great semicircles. But opposite points on the boundary must still be identified. This version of the projective plane is therefore a topological distortion of figure 11.4.*

"And, topologically speaking, a hemisphere *is* a disc. It's just got a bit bent. So the new model is just the old one in disguise."

"Bravo!" They applauded. I suspected irony, but I played along and gave a low bow.

"*Bis!*" yelled Elena, meaning "encore", getting carried away by the spirit of things. Enrico tried to shut her up but the damage was done. I began to perform my encore.

"There are lots of different ways to visualize the projective plane," I said. "Dozens." (Enrico groaned.) "One of them was discovered by Jacob Steiner. Well, sort of. The year was 1844, and by coincidence he was visiting Rome, so he called it the *Roman surface* (figure 11.10). It's one of the few mathematical objects named after a *place*. In actual fact, he constructed it in a highly complicated fashion using pure geometry. Now, using coordinates, every surface is determined by some equation. For instance, a sphere of radius 1 centred at the origin has equation $x^2 + y^2 + z^2 = 1$ in coordinates (x,y,z). Steiner was a wonderful geometer but hopeless at algebra, and he couldn't work out the equation for his surface. A year before Steiner died he asked Karl Weierstrass to work the equation out. Weierstrass, a much more versatile mathematician than Steiner, found the equation with no trouble at all:

$$x^2y^2 + y^2z^2 + z^2x^2 + xyz = 0.$$

It's beautifully symmetric, just like the surface."

Enrico and Elena admired the elegant symmetry of their native city's surface.

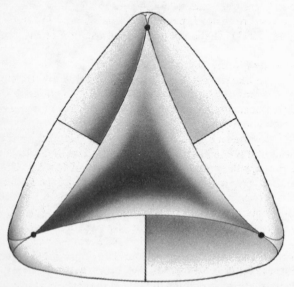

11.10   *Steiner's Roman surface: six cross-caps joined together. It has the same symmetry as a tetrahedron.*

"The Klein bottle has an equation, too," I said. "It's formed by the points (x,y,z) such that

$$(x^2 + y^2 + z^2 + 2y - 1)\left((x^2 + y^2 + z^2 - 2y - 1)^2 - 8z^2\right) +$$
$$16xz(x^2 + y^2 + z^2 - 2y - 1) = 0.$$

It's not as symmetric; but then neither is the surface.

"The Roman surface is just a projective plane in yet another disguise. But it has a flaw." They shook their heads in horror at this news. "Like the cross-cap, it has (several) *singular points*. Those are places where it doesn't just cross through itself in two or more separate sheets, but the sheets get all tangled up and merge together. Like the top of the cross-cap. The Klein bottle, on the other hand, has no singular points. It crosses itself, but in clearly defined separate sheets. Maybe that's why most people don't realize that the projective plane is really a *simpler* example of a one-sided surface. It's a lot easier to draw a convincing Klein bottle."

"It's a snappier name, too. Someone did a better public relations job."

"Could be. For a long time it was actually an unsolved question whether the projective plane can be arranged in three-dimensional space so that it has no singular points – only self-intersections. In fact David Hilbert, one of the greatest mathematicians who ever lived, conjectured that it can't be done – and told his student Werner Boy to prove it. Boy, like any good research student, followed his own nose and *disproved* Hilbert's conjecture instead, producing what is now known as *Boy's surface* (figure 11.11). That's yet another incarnation of the projective plane."

"It looks sort of funny," said Elena.

"Yes. It's a bit like three doughnuts stuck together, but the dough of each doughnut runs into the hole in the next. There's a polyhedral model for the Boy surface, which you can make from cardboard (figure 11.12). That may give you a better idea of the shape."

"Does the Boy surface have a pretty equation, like Steiner's?" asked Elena. I thought it was a highly intelligent question, and told her so.

"That's perhaps the greatest curiosity of them all," I said. "Until very recently, nobody knew the answer. They could draw the surface, they could study its topology, but they couldn't decide whether or not it had a polynomial equation, pretty or not. In 1978 Bernard Morin, a French geometer who, incidentally, is blind, found equations for a projective plane without singularities but nobody could prove it was the same as the Boy surface. In 1985 J. F. Hughes found an empirical formula using polynomials of degree 8. But both formulas are *parametrizations*; that is, instead of an equation 'something in $x,y,z = 0$' they take the form '$x, y,$ and z = certain expressions in some other variables'. In principle you can

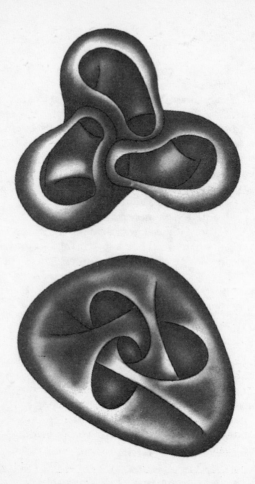

11.11   *Boy's surface, topologically equivalent to the projective plane and having no singular points. Hilbert conjectured that no such surface exists. The self-intersections (solid lines) form a "bouquet" of three loops joined at a common point. These two very different views are topologically equivalent. In each, sections of the surface have been cut away to reveal the interior.*

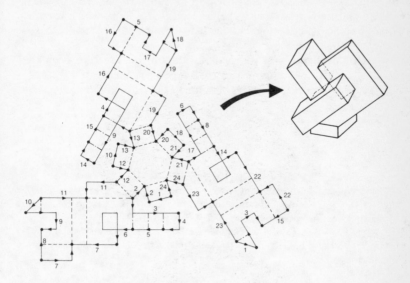

11.12 *To make a polyhedral model of Boy's surface, cut this shape from thin card and join the edges with the same numbers.*

eliminate the new variables and get some hugely complicated equation in x, y, and z, but I don't think anybody has done it.

"In 1986 François Apéry found an explicit equation for the Boy surface (box 11.1), a polynomial of the sixth degree. It was derived by deforming Steiner's Roman surface to get rid of its singularities. Sounds like a simple trick, but nobody had been able to make it work before."

"Amazing," said Enrico.

Box 11.1   François Apéry's Equation for the Boy surface

$$64(1-z)^3 z^3 - 48(1-z)^2 z^2 \left(3x^2 + 3y^2 + 2z^2\right) +$$

$$12(1-z)z\left[27\left(x^2+y^2\right)^2 - 24z^2\left(x^2+y^2\right) + 36\sqrt{2}yz\left(y^2-3x^2\right)\right.$$

$$\left. + 4z^4\right] + \left(9x^2 + 9y^2 - 2z^2\right)\left[-81\left(x^2+y^2\right)^2 - 72z^2\left(x^2+y^2\right)\right.$$

$$\left. + 108\sqrt{2}xz\left(x^2-3y^2\right) + 4z^4\right] = 0$$

"There's a lot more," I said. "If you've got time, I can tell you how the Boy surface can be used to turn a sphere inside out. The story involves Morin, and a French physicist called Jean-Pierre Petit, which is really rather a coincidence seeing that 'klein' in German means 'small', and so does 'petit' in . . ."

But my audience was disappearing down the Via XX Settembre, heading rapidly and determinedly towards the point at infinity . . .

Next time I meet them, I'm going to tell them about *finite* projective planes.

FURTHER READING

François Apéry, "La Surface de Boy", *Advances in Mathematics*, 61 (1986)

François Apéry, *Models of the Real Projective Plane* (Braunschweig: Vieweg, 1987)

Werner Boy, "Über die Curvatura integra und die Topologia geschlossener Flächen", *Mathematische Annalen*, 57 (1903), pp. 151–84

George K. Francis, *A Topological Picture-book* (New York: Springer-Verlag, 1987)

David Hilbert and S. Cohn-Vossen, *Geometry and the Imagination* (New York: Chelsea, 1983)

12

The Twelve Games of Christmas

It was Christmas afternoon at Baffleham Hall. The turkey was long dismembered, the Christmas pudding devoured. Lord Roderick of Baffleham and his eleven guests were in a relaxed mood, singing the traditional songs . . .

. . . Three Frinch hins,
Two titled doves,
And a *pah*tri-idge in a *peer*-tree.

The vigorous delivery compensated for the imperfections of pitch and the upper-class accent.

"Oh! Zat is most curieux," exclaimed the Comtesse de Malfamée. "I 'ave never 'ear 'im as a Christmas song. But in la belle France we 'ave a traditional song like zat, about ze monz of ze year . . .

> Au premier mois de l'année,
> Que donn'rai-je à ma mie?
> Une perdriole,
> Que va, que vient, que vole
> Une perdriole
> Que vole dans le vent.

"And he ends wiz:

Au douzième mois de l'année,	On the twelfth month of the year,
Que donn'rai-je à ma mie?	What shall I give to my lady-love?
Douze coqs chantant,	Twelve singing cocks,
Onze ortolans,	Eleven ortolans [buntings],
Dix pigeons blancs,	Ten white pigeons,
Neuf boeufs cornus,	Nine horned bulls,
Huit moutons tondus,	Eight sheared sheep,
Sept chiens courants,	Seven running dogs,
Six lièvres aux champs,	Six hares in the fields,
Cinq lapins courant par terre,	Five rabbits running on the ground,
Quat' canards volant en l'air,	Four ducks flying in the air,
Trois ramiers de bois,	Three woodpigeons,
Deux tourterelles,	Two turtle doves,
Une perdriole,	A partridge,
Que va, que vient, que vole	Who goes, who comes, who flies,
Une perdriole	A partridge,
Que vole dans le vent."	Who flies in the wind.

The others applauded enthusiastically, and the Comtesse gave a little curtsy. "We call zis song *La Perdriole*."

"I take it a perdriole is a partridge," said the Duke of Balmuddle.

"Mais oui," said the Comtesse.

"But no pear-tree. How extraordinarily odd," chimed in the Duke's second son, Edmund. "There must be some relationship . . . " He hauled out a copy of the *Oxford English Dictionary*. "Partridge. Middle English *pertrich, partrich* . . . from Old French *perdriz, pertris* . . . My word! That sounds just like pear-tree! Look, father – the Old French word for 'partridge' is pronounced 'pear-tree'!"

"Sounds to me," said Lady Chattermere, "as if someone got the words confused. As I recall, there's some question about what 'calling birds' are . . . Or whether it was 'colly birds', whatever *those* might be."

"Perhaps it should be four collie dogs," said Annabel, her daughter, and giggled.

"'Colly' . . . Obsolete word for 'blackbird'," said Edmund, thumbing through the dictionary.

"Oh, shut up, Edmund!" said Annabel's annoying small brother, Charles. "Don't be such a bighead!"

"What I've never understood," said the elderly Baron Goutsfoot, "is why the damn' song . . ."

"*Grandfather!* Language!"

"Sorry, Hilda m'dear . . . Why the confounded song starts with birds and animals and suchlike, but ends up with lords and ladies and drummers. Damn' inconsistent, if you ask me. Beggin' your pardon, Hilda m'dear."

"Ze words to ze French song differ from one region to anuzzer," said the Comtesse's daughter, Esmeralde, who was studying poetry at the Sorbonne. "In ze earliest version, from ze sixteenz century, it begins wiz 'Douze chevaliers, onze demoiselles . . .' but zen it goes back to animals. And zere is a Canadian version wiz ze *days* of ze year, which goes on and on and on . . . Zey sing it to put children to sleep."

"I've always wondered," put in Orville, Hilda's current beau, up from Oxford for the weekend, "what the five gold rings were for."

"To string through your nose, darling," said Hilda.

"Children, children," said Lord Roderick. "Cease your bickering! For now is the time for the oldest tradition of Baffleham Hall – the Christmas puzzles! I hope you've all come prepared?" There were nods all round the table, except for Uncle Crispin, who had dozed off. "And, seeing as there are twelve of us, I suggest we try to play a puzzle version of the *Twelve Days of Christmas*! And I warn you, by long tradition, at least one puzzle is *really* hard . . . But of course I'm not saying which.

"You first, Edmund."

Twelve drummers drumming . . .

"Gosh, thanks, Roddy . . ." said Edmund, who had prepared a problem about card players. "Um . . . Well, this puzzle is about – er – twelve drummers." He paused.

"Go on, Edmund."

"Er. Gosh . . . Well, you may not know this, but I used to be in the Guards."

"Yes, he was head guarderer," said Hilda sweetly.

"No, the Relief Airborne Footguards. You must have heard of the RAF! Anyway, I was in the regimental band . . ."

"The regiment was OK, it was just Edmund who was banned . . ."

"No, he was an aide to the Drum Major. They called him the band-aide."

"Thank you, Annabel. But, speaking of drums, there was a regimental tradition, the drumming contest. There were several events, actually; the singles, the doubles . . ."

"And the mixed doubles."

"You may scoff. The doubles were a sort of league. There were twelve drummers altogether and they formed up into pairs. The pairs weren't fixed, of course: the idea was to find the best pair. The contest lasted eleven days . . ."

"Well, there's a lot of variety in drumming, got to give people a chance to show what they can . . ."

"No, we didn't play *all* day. There was just one round every day, held before reveille so as not to disturb anyone. You see, with twelve drummers there are sixty-six different pairs. Each round consisted of three separate contests between two pairs of drummers, because of course that makes twelve drummers altogether. So that's six pairs per day, and six into sixty-six goes eleven times."

Annabel clapped with heavy irony.

"On any given day all twelve drummers took part. And, to make it fair, every drummer had every other drummer exactly once as a partner, and exactly twice as an opponent.

"The puzzle is . . ."

" . . . why on earth they bothered."

"No, Charles, and take your paws out of the trifle. No, the puzzle is: how did they do it?"

Eleven pipers piping . . .

"You probably won't know this," Hilda began. "But, like Edmund, I have a musical past. I used to be a flautist in the London Philharmonic Orchestra."

"Actually," said Annabel in a stage whisper, "it was the Dagenham Girl Pipers, and she flauted everything she had."

Hilda sniffed. "I left after a dispute. They treated me *most* unfairly. There were eleven flautists, you see, and one day a consignment of new flutes arrived. The first flautist took one eleventh of the flutes plus one eleventh of a flute . . ."

"How on earth can you play one eleventh of a flute?"

"Well, no flutes actually got cut up, you understand. The next took one tenth of the remainder, plus one tenth of a flute. Then the next took one ninth of the remainder plus one ninth of a flute, and so on in turn . . . The next to last took half of what remained plus half a flute. I was the last. When I saw how many they'd left me, I got annoyed, and resigned on the spot."

"Why?"

"Everyone else got twice as many flutes as me."

"How *terrible*! But what's the puzzle, dear?"

"How many flutes were there in the consignment?"

Ten lords a-leaping . . .

Baron Goutsfoot levered himself out of a Queen Anne chair and grumped his way to the writing-desk, where he borrowed a piece of paper and sketched a diagram (figure 12.1). It showed twenty-eight circles joined by lines to form a triangular network. He rummaged through his pockets, found ten gold sovereigns, and placed them on the central ten dots.

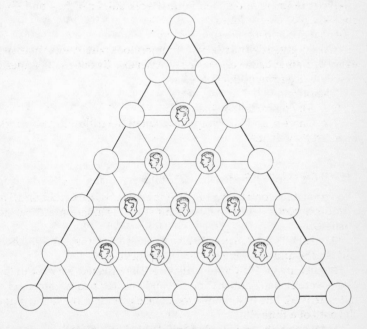

12.1   *How can ten lords leap, leaving only one?*

"This game symbolizes how decisions are taken at the highest political levels," he grunted. "The triangle represents the House of Lords, which as we know *really* runs the country, and the circle in the middle (number 13, an auspicious number) stands for a seat on the Privy Council."

"Respectfully, Baron," said Orville, "I must point out that it is not possible to stand for a seat. One may sit for a seat or stand for a –uh–*stand*, but . . ."

"Nonsense! Cousin Dominic stood for a seat in Upward-le-Mobile! Lost it to some SMP oick, rotten bad luck, just because the gutter press found out about him wanting to sell the common to a cruise missile manufacturer . . ."

"Thank you, Annabel. Hmmph. The ten sovereigns represent ten lords. They take it in turns to leap over any adjacent lord into a vacant circle immediately beyond in a straight line. The lord who is jumped over has lost influence and is removed. The first puzzle is, how can they do this to end up with a single lord sitting on the Privy Council?"

"It's an awfully complicated way to sit on the privy!" yelled Charles, and was cuffed for his pains.

"Golly," said Edmund. "It's just like solitaire."

"A *bit* like solitaire," admitted Baron Goutsfoot in a hurt tone of voice. "Edmund, since you're so clever, I'll give *you* a *different* puzzle. Three of the lords want to become Law Lords, who sit in the outer corners of the House. How can the ten lords start in the same position, jump over each other following the same rules, and end up with *three* lords, one in each of the three corners (numbers 1, 22, 28)?"

"Is this the really hard one?" asked Edmund worriedly.

"Might be. Might not."

Nine ladies dancing . . .

"This one's mine!" exclaimed the lecherous but ageing Duke of Balmuddle, whose aspirations exceeded his capabilities. "Bring on the dancing-girls!"

"Remember, Courtney, that there is a minor present," warned Lady Chattermere. "In any case, you are too old for dancing-girls."

"I'm not a minor, Uncle Courtney!" yelled Charles. "They don't send children down the mines any more!"

"I assure you, madam, that it will be perfectly respectable," said the Duke.

"Bother!" whispered Charles.

"In Castle Balmuddle there is a dance which goes back to the days of Clan MacCroney, when the Great Cameron Dunrovin himself took the pipes. Ah, those were great days for Scotland! 'Tis called the Fichin Reel.

Nine ladies arrange themselves in a circle. Three wear green bonnets, three red bonnets, and three blue bonnets.

"The ladies take it in turns to dance in pairs in the circle while the others twirl on the spot. After four pairs have danced, the remaining lady dances in the circle on her own.

"The pattern of the dance is that the first pair determine the order of all the others, in this manner. If you remove a pair of ladies from a circle you create two arcs of adjoining ladies – unless of course the original two are adjacent, in which case only one arc is left. If a pair of ladies with the same colour bonnet dance, then the next pair is formed by the ladies at the ends of the longer arc left in the circle. If a pair of ladies with different coloured bonnets dance, then the next pair is formed from the two ladies at the ends of the shorter arc. If an arc with just one lady in it is created, she must dance alone.

"The first pair is chosen freely, but thereafter the stated rules apply. The problem is how to arrange the ladies so that all nine dance, four in pairs and one alone. I must add that Cameron Dunrovin himself decreed that no three ladies with identical coloured bonnets may stand next to each other in the circle, but that somewhere in the circle two ladies with red bonnets must stand together, two with blue, and two with green."

Eight maids a-milking . . .

"In ze milkshed," explained Esmeralde, "zere were eight maids, milking eight cows, sitting in a circle. Clockwise round ze circle ze maids 'ad buckets zat could 'old exactly 3, 4, 5, 6, 7, 8, 9, 10 – 'ow you say, gallons? – of milk. When zey 'ave finish milking, buckets 4, 5, 6, and 9 are full, and ze udders are empty."

"Well, they would be if the maids had finished," said Edmund. "Oh! Wait, you mean the *others*!"

"Zat is what I say, no?" said Esmeralde, perplexed. "Now ze maids must return to ze farm wiz ze same amount each, zat is, three gallons in each bucket. Uzzerwise ze farmer 'e will be vair' angry zat some of ze maids are lazy.

"Zey can pour milk from any bucket into zose adjacent around ze circle. So 'ow can ze milkmaids share out ze milk, so zat each bucket contain exactly three gallons?"

Seven swans a-swimming . . .

"Well," said Orville, "I hadn't actually planned to do one about *swans*, you know . . . Had a real mindboggler about *ducks*, though . . ."

"Shut up, Orville!" screamed Charles. "It's my turn anyway! You're supposed to do the French hens!" He looked around him at the assembled company. "Though why it's *three* French hens I can't imagine," he said. The Comtesse and her daughter coloured but pretended not to notice the insult.

"You may have your turn, Charles," said Lady Chattermere. "And if you say another word out of place you will spend the rest of Christmas in your room."

"Yes, mother. Sorry, mother. Now, there were seven swans who lived in eight lakes connected by canals, like this (figure 12.2). They liked to sun themselves on the banks, but when the gamekeeper came to feed them they would all jump into the water.

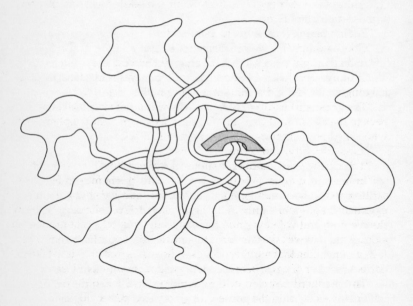

12.2   *Swan lakes.*

"One after the other, each swan jumped into an empty lake and then swam along a canal to the next lake, which was also empty.

"How did they do it?"

"That's a stupid puzzle," said Orville. "For a start, the canals cross."

"Those are aqueducks," said Charles.

"Aque*ducts*, you little squirt. Anyway, you just start *here*, and move the first swan under the bridge to . . ."

"But there was a poacher hiding under the bridge," said Charles. "And the swans knew he'd catch them and eat them for Christmas dinner, so they didn't go that way."

Six geese a-laying . . .

With a lot of effort they managed to wake up Uncle Crispin and explain to him the rules of the Twelve Games of Christmas.

"Where have we got to, then?"

"Six geese a-laying."

"Ah, geese . . . Six geese . . . Right! This is a really hard one, all about a goose-girl called Bumps."

"Zat is a strange name," said Esmeralde.

"Goose-bumps," explained Crispin.

"In all the right places, I bet," said Orville with a leer.

"Shut up, Orville!" said Charles. Orville raised his fist; Charles stuck his tongue out and hid behind a suit of armour.

"The goose-girl had six geese, and she was taking her basket of eggs back to the farm," said Crispin. "On the way she met the shepherd-boy, who asked how many eggs her geese had laid."

"'Less than fifty, and prime,' said she."

"'I mean the exact number,' said the shepherd-boy."

"In answer, the girl produced a cube, with the names of her geese written on the faces, numbered just like an ordinary die but with goose-eggs instead of spots (figure 12.3). 'We'll play a game,' she said. 'You roll the die first, and whatever number comes up, you take that number of eggs. After that we take it in turns to give the die a quarter turn to a new face, again taking that number of eggs. Whoever is first unable to take the correct number of eggs, because there are not enough left, loses.'"

"The shepherd-boy said he'd understood, and rolled the die."

"'You've lost,' said the goose-girl, who was a perfect logician."

"The shepherd-boy said that he should have thrown one higher – or maybe one lower."

"'You'd still have lost,' she told him. 'But by proper play you could have won if you'd thrown anything else.' "

Crispin stopped, a smile on his face, and sat down again.

"Is that it?" asked Annabel.

"What's the puzzle?" said the Duke of Balmuddle.

"Woops, forgot that bit," said Crispin. "How many eggs were there in the basket, and what did the shepherd-boy throw?"

12.3   *The goose-girl's die, with eggs for spots.*

Five gold rings . . .

"My turn now, Roderick," said Lady Chattermere firmly. She delved into her purse and brought forth a strange piece of jewellery. "Heirloom, been in the family since the days of Lord Jocelyn, who was carrying five gold rings and twenty diamonds in his pack when he was hit in the chest by shrapnel during the Battle of Baghdad ."

"Hit in the chest? That sounds nasty."

"It was! It was the chest containing the regimental pay!"

"Which consisted of five gold rings and twenty diamonds, perhaps?"

"Good Lord, no! There are plenty of ways to fiddle regimental accounts without actually appropriating goods! A clever quartermaster can make a for- . . . But that isn't really important. Anyway, when the battle was over Lord Jocelyn wanted the rings to be fashioned into a . . . sort of brooch. When the jeweller heard the instructions, he thought maybe Lord Jocelyn had been hit in the *head*.

"As you can see, there are five gold rings.

"'I want all five rings to be joined together,' Lord Jocelyn told him. 'Like *this* (figure 12.4), and set with diamonds where they cross.'

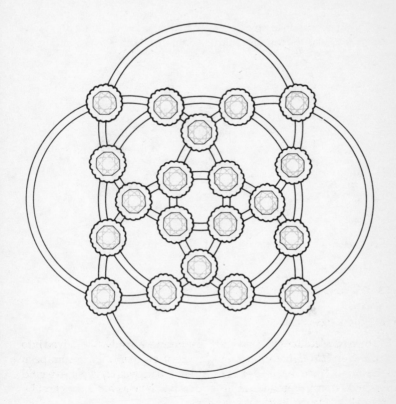

12.4 *Five gold rings.*

  "'Easy,' said the jeweller. 'I mean – that will be *very* expensive, but I'm sure I can do a perfect job.'

  "'Good,' said Lord Jocelyn. 'But there's an extra condition. There are twenty diamonds altogether, and it so happens that their weights are 1 carat, 2 carats, . . . and so on up to 20 carats. There are eight diamonds on each ring. They must be arranged so that the total number of carats on each ring is the same.'

  "The jeweller managed, eventually, hence the existence of the heirloom. But can *you* think of a suitable arrangement?"

Four calling birds . . .

"As Edmund explained so grippingly," said Annabel, "they are in fact four *colly* birds, that is, blackbirds. Four blackbirds were sitting on mushrooms." (Figure 12.5)

12.5   Four colly birds.

"Two of them are white, Annabel," Edmund pointed out.

"Two blackbirds and two albino blackbirds were sitting on mushrooms," Annabel continued.

"So called because there isn't mush room for a bird to sit on them," put in Charles.

"Quiet, brat. You've heard of a fairy ring? Well, this fairy ring consisted of eight mushrooms. The two blackbirds and the two albinos were sitting on every alternate mushroom, with the blackbirds to the north and the albinos to the south.

"They wanted to change places so that the two blackbirds were on the mushrooms occupied by the albinos, and vice versa. But to avoid offending the fairies who had built the ring, they were only allowed to do this by moving round exactly three mushrooms at a time, either clockwise or anticlockwise. Three being a magic number, you understand.

"How did they do it?"

Three French hens . . .

"This puzzle concerns triplets called Nicole, Nathalie, and Nancy."

"It's supposed to be about *hens*," said Charles.

"Three French N's," Orville explained.

"Nancy isn't a French name!"

"Yes it is, there's a town called Nancy just west of Strasbourg. These young ladies obeyed one immutable rule. Nicole alternately lied and told the truth, Nathalie always told the truth, and Nancy always lied. Unfortunately they all looked exactly the same and no one could ever tell them apart.

"One day the three French N's were sitting next to each other on the step of their house . . ."

"The N-house, no doubt," said Charles in a tone of disgust.

"The following conversation ensued. Here R is the right-hand lady, M the middle lady, and L the left-hand lady.

L [to M]:	You're a liar.
M:	No I'm not!
R:	You're both liars.
L:	That was a lie!
M:	That was a lie!
R:	That was a lie!

Which French N is which?"

Two turtle doves . . .

"It is a little known fact," said the Comtesse de Malfamée, "zat some birds is vair' good at arizmetic. One day zere was two turtle dove sitting on a branch. 'I am sinking of a number under one 'undred,' coo one dove. 'Me too,' reply ze uzzer. 'Tell me your number,' 'e adds. She does. 'E tell 'er 'is, remarking as 'e do so zat 'is number and 'ers 'ave no digits in common. 'Coo, zat is remarkable,' she say. 'If we add our numbers and square ze result, zen we get a four-digit number ooze first two digits are your number and ze last two digits are mine.' 'You mean, like 30 plus 25 is 55

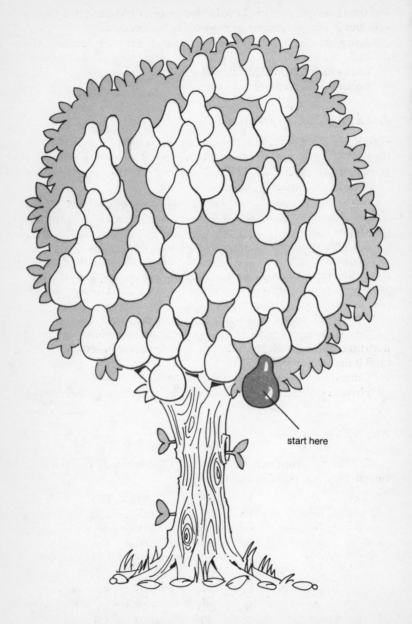

start here

12.6   *Which pear is the partridge?*

and if zat is square zen it make 3025?' ask ze gentilhomme dove. 'Oui,' she say, 'but not zose numbers.' Zo what *was* ze numbers?"

"I suppose," the Comtesse added, "zat is why zey are call turtle doves."

"Eh?" asked Lord Roderick, perplexed.

"Because zey calculate ze turtle before zey square it!"

And a partridge in a pear-tree . . .

Lord Roderick of Baffleham put down his cigar and got to his feet. "It is the host's prerogative to bring this traditional event to a close. In the garden of Baffleham Hall," he said ponderously, "there is a pear-tree."

"Hundreds, I imagine, sir" said Orville.

"Hrrumph, yes. Well, I do have one particular pear-tree in mind, Orville. Now one of the pears in this tree (figure 12.6) is actually a partridge."

"Oh, jolly good, sir! Fat little blighter with no legs, is it?"

"The partridge, Orville, is partially concealed by neighbouring pears. It is hiding from a hunter. Now, to get into the tree, the partridge started at the large pear dangling from the lowest branch, and then passed from pear to pear in some order."

"Ah," said Crispin. "It's a puzzle about ordered pears."

"Yes, Crispin, so good of you to point that out. Now in fact, the partridge crossed each boundary between adjacent pears exactly once, until it finally hid itself."

"Cunning little devil! Dashed good show, sir!"

"Hrrumph. Well, what I want to know is . . . *which pear is the partridge?*"

ANSWERS

12. Let the drummers be denoted by the letters A–L. Then on the eleven days, one possible arrangement is:

AB – IL	EJ – GK	FH – CD
AC – JB	FK – HL	GI – DE
AD – KC	GL – IB	HJ – EF
AE – LD	HB – JC	IK – FG
AF – BE	IC – KD	JL – GH
AG – CF	JD – LE	KB – HI
AH – DG	KE – BF	LC – IJ
AI – EH	LF – CG	BD – JK
AJ – FI	BG – DH	CE – KL
AK – GJ	CH – EI	DF – LB
AL – HK	DI – FJ	EG – BC

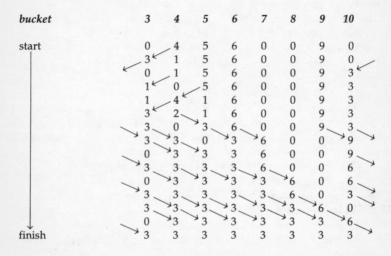

11.   Twenty-one. Everyone else took two, Hilda got one.

10.   Number the cells as in figure 12.7. Then the answers are:
(a) $13 \rightarrow 4, 17 \rightarrow 8, 14 \rightarrow 25, 25 \rightarrow 12, 5 \rightarrow 14, 20 \rightarrow 9, 12 \rightarrow 5, 4 \rightarrow 6, 6 \rightarrow 13.$
(b)   Impossible. Colour cells 1, 4, 6, 11, 13, 15, 22, 24, 26, 28. Then anything starting on a coloured circle finishes on one, and vice versa. The three corner cells are coloured – but only one lord, the central one, starts on a coloured circle.

9.   G G B B R R G R B, starting at G G. There are others.

8.   Here's one possibility. The arrows show the movement of milk between adjacent buckets.

bucket	3	4	5	6	7	8	9	10
start	0	4	5	6	0	0	9	0
	3	1	5	6	0	0	9	0
	0	1	5	6	0	0	9	3
	1	0	5	6	0	0	9	3
	1	4	1	6	0	0	9	3
	3	2	1	6	0	0	9	3
	3	0	3	6	0	0	9	3
	3	3	0	3	6	0	0	9
	0	3	3	3	6	0	0	9
	3	3	3	3	6	0	0	6
	0	3	3	3	3	6	0	6
	3	3	3	3	3	6	0	3
	3	3	3	3	3	3	6	0
	0	3	3	3	3	3	3	6
finish	3	3	3	3	3	3	3	3

7.   Number the lakes clockwise from 1 to 8 as in figure 12.8. Then the successive moves are $6 \rightarrow 1, 3 \rightarrow 6, 8 \rightarrow 3, 5 \rightarrow 8, 2 \rightarrow 5, 7 \rightarrow 2, 4 \rightarrow 7.$ Each swan swims towards the lake that the previous one jumped into.

6.   Forty-three eggs, and he threw a 3. This is the "very difficult" puzzle that Lord Roderick promised. It depends on a general analysis of the winning strategy, when the total to aim for is arbitrary. The strategy can be found by working back from the end position, and is shown in the next table: this gives the winning move(s) for each combination of current total and face of the die. "L" means a losing position – whatever move you make, the opponent can then win with perfect play.

The result is a pattern that repeats when the total increases by 9, with a few exceptions at the beginning. Thus, for totals larger than 7, the strategy depends only on the digital root of the total (the sum of the digits, repeated until the answer is in the range 1–9). The goose-girl, being a

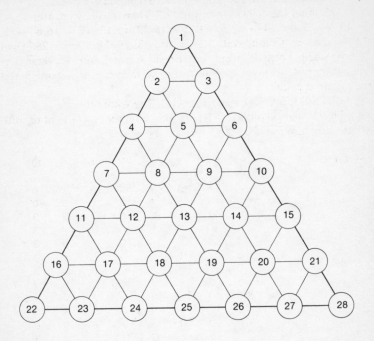

12.7   *Ten lords a-leaping.*

	face showing		
total	1 or 6	2 or 5	3 or 4
1	L	1	1
2	2	1	1,2
3	3	3	L
4	4	4	L
5	5	L	5
6	3	6,3	6
7	2,3,4	6,3,4	6,2
8	4	4	L
9	L	L	L
10	5	1	1,5
11	2,3	3	2
12	3,4	3,4	L
13	4	4	L
14	5	L	5
15	3	6,3	6
16	2,3,4	3,4	2

17 onwards: repeat from 8

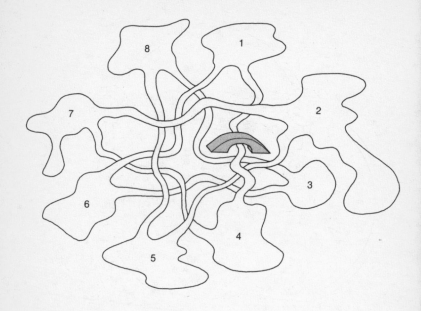

12.8   *Seven swans a-swimming.*

perfect logician, worked this pattern out instantly. We are told that the boy's throw, the number above it, and the one below all lead to a losing position. The strategy pattern shows that the only three consecutive throws that can lead to a losing position are 2, 3, 4, so the boy threw the middle one: 3. The numbers of eggs for which exactly 2, 3, 4 are losing throws are those with a digital root of 7, with the exception of 7 itself, for which 6 is also a losing throw. The only other prime less than 50 with digital root 7 is 43. So there were forty-three eggs.

  5.   See figure 12.9.

  4.   If the mushrooms are numbered as in figure 12.5, one solution is: $8 \rightarrow 3, 2 \rightarrow 5, 5 \rightarrow 8, 4 \rightarrow 7, 7 \rightarrow 2, 2 \rightarrow 5, 6 \rightarrow 1, 1 \rightarrow 4, 4 \rightarrow 7, 7 \rightarrow 2, 3 \rightarrow 6, 6 \rightarrow 1, 1 \rightarrow 4, 8 \rightarrow 3, 3 \rightarrow 6, 5 \rightarrow 8.$

  3.   There are six possible arrangements. By trial and error, we find that Nathalie is on the left, Nancy is in the middle, and Nicole on the right.

  2.   98 and 01.

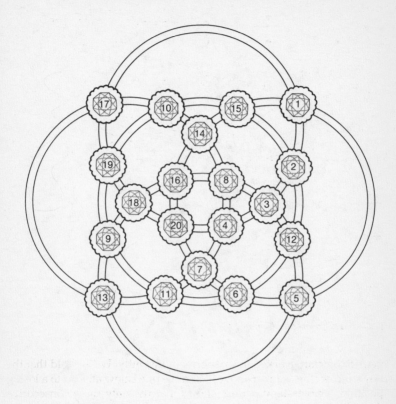

12.9 *Five gold rings plus diamonds.*

1.   Whenever the partridge enters and leaves the space occupied by a pear, it crosses two boundaries between adjacent pears. Hence, except at the ends of the journey, each pear must touch an even number of other pears. Therefore the pears at the ends of the journey are those that touch an odd number. The pear from which the partridge starts touches one other pear. The only pear in the tree, apart from this, that touches an odd number of pears, is the one shown in figure 12.10, with three neighbours. The figure also shows one possible route.

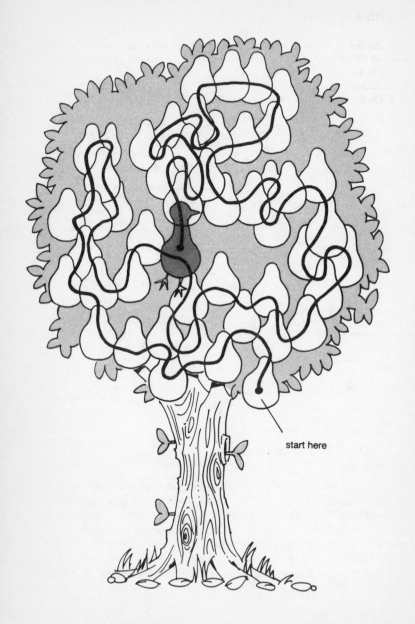

start here

12.10 *The partridge in the pear-tree.*

FURTHER READING

W. S. Andrews, *Magic Squares and Cubes* (New York: Dover Publications, 1960)

Martine David and Anne-Marie Delrieu, *Aux Sources des chansons populaires* (Paris: Belin, 1982)

H. E. Dudeney, *Amusements in Mathematics* (New York: Dover Publications, 1958)

H. E. Dudeney, *The Canterbury Puzzles* (New York: Dover Publications, 1958)

FOR THE BEST IN PAPERBACKS, LOOK FOR THE 🐧

In every corner of the world, on every subject under the sun, Penguin represents quality and variety – the very best in publishing today.

For complete information about books available from Penguin – including Puffins, Penguin Classics and Arkana – and how to order them, write to us at the appropriate address below. Please note that for copyright reasons the selection of books varies from country to country.

In the United Kingdom: Please write to *Dept E.P., Penguin Books Ltd, Harmondsworth, Middlesex, UB7 0DA.*

If you have any difficulty in obtaining a title, please send your order with the correct money, plus ten per cent for postage and packaging, to *PO Box No 11, West Drayton, Middlesex*

In the United States: Please write to *Dept BA, Penguin, 299 Murray Hill Parkway, East Rutherford, New Jersey 07073*

In Canada: Please write to *Penguin Books Canada Ltd, 2801 John Street, Markham, Ontario L3R 1B4*

In Australia: Please write to the *Marketing Department, Penguin Books Australia Ltd, P.O. Box 257, Ringwood, Victoria 3134*

In New Zealand: Please write to the *Marketing Department, Penguin Books (NZ) Ltd, Private Bag, Takapuna, Auckland 9*

In India: Please write to *Penguin Overseas Ltd, 706 Eros Apartments, 56 Nehru Place, New Delhi, 110019*

In the Netherlands: Please write to *Penguin Books Netherlands B.V., Postbus 195, NL–1380AD Weesp*

In West Germany: Please write to *Penguin Books Ltd, Friedrichstrasse 10–12, D–6000 Frankfurt/Main 1*

In Spain: Please write to *Alhambra Longman S.A., Fernandez de la Hoz 9, E–28010 Madrid*

In Italy: Please write to *Penguin Italia s.r.l., Via Como 4, I-20096 Pioltello (Milano)*

In France: Please write to *Penguin Books Ltd, 39 Rue de Montmorency, F-75003 Paris*

In Japan: Please write to *Longman Penguin Japan Co Ltd, Yamaguchi Building, 2–12–9 Kanda Jimbocho, Chiyoda-Ku, Tokyo 101*

FOR THE BEST IN PAPERBACKS, LOOK FOR THE 🐧

PENGUIN PHILOSOPHY

I: The Philosophy and Psychology of Personal Identity   Jonathan Glover

From cases of split brains and multiple personalities to the importance of memory and recognition by others, the author of *Causing Death and Saving Lives* tackles the vexed questions of personal identity. 'Fascinating ... the ideas which Glover pours forth in profusion deserve more detailed consideration' – Anthony Storr

Minds, Brains and Science   John Searle

Based on Professor Searle's acclaimed series of Reith Lectures, *Minds, Brains and Science* is 'punchy and engaging ... a timely exposé of those woolly-minded computer-lovers who believe that computers can think, and indeed that the human mind is just a biological computer' – *The Times Literary Supplement*

Ethics   Inventing Right and Wrong   J. L. Mackie

Widely used as a text, Mackie's complete and clear treatise on moral theory deals with the status and content of ethics, sketches a practical moral system and examines the frontiers at which ethics touches psychology, theology, law and politics.

The Penguin History of Western Philosophy   D. W. Hamlyn

'Well-crafted and readable ... neither laden with footnotes nor weighed down with technical language ... a general guide to three millennia of philosophizing in the West' – *The Times Literary Supplement*

Science and Philosophy: Past and Present   Derek Gjertsen

Philosophy and science, once intimately connected, are today often seen as widely different disciplines. Ranging from Aristotle to Einstein, from quantum theory to renaissance magic, Confucius and parapsychology, this penetrating and original study shows such a view to be both naive and ill-informed.

The Problem of Knowledge   A. J. Ayer

How do you *know* that this is a book? How do you *know* that you know? In *The Problem of Knowledge* A. J. Ayer presented the sceptic's arguments as forcefully as possible, investigating the extent to which they can be met. 'Thorough ... penetrating, vigorous ... readable and manageable' – *Spectator*

FOR THE BEST IN PAPERBACKS, LOOK FOR THE 🐧

PENGUIN BUSINESS AND ECONOMICS

Almost Everyone's Guide to Economics   J. K. Galbraith and Nicole Salinger

This instructive and entertaining dialogue provides a step-by-step explanation of 'the state of economics in general and the reasons for its present failure in particular in simple, accurate language that everyone could understand and that a perverse few might conceivably enjoy'.

The Rise and Fall of Monetarism   David Smith

Now that even Conservatives have consigned monetarism to the scrapheap of history, David Smith draws out the unhappy lessons of a fundamentally flawed economic experiment, driven by a doctrine that for years had been regarded as outmoded and irrelevant.

Atlas of Management Thinking   Edward de Bono

This fascinating book provides a vital repertoire of non-verbal images that will help activate the right side of any manager's brain.

The Economist Economics   Rupert Pennant-Rea and Clive Crook

Based on a series of 'briefs' published in *The Economist*, this is a clear and accessible guide to the key issues of today's economics for the general reader.

Understanding Organizations   Charles B. Handy

Of practical as well as theoretical interest, this book shows how general concepts can help solve specific organizational problems.

The Winning Streak   Walter Goldsmith and David Clutterbuck

A brilliant analysis of what Britain's best-run and most successful companies have in common – a must for all managers.

FOR THE BEST IN PAPERBACKS, LOOK FOR THE 🐧

PENGUIN SCIENCE AND MATHEMATICS

Facts from Figures M. J. Moroney

Starting from the very first principles of the laws of chance, this authoritative 'conducted tour of the statistician's workshop' provides an essential introduction to the major techniques and concepts used in statistics today.

God and the New Physics Paul Davies

Can science, now come of age, offer a surer path to God than religion? This 'very interesting' (*New Scientist*) book suggests it can.

Descartes' Dream Philip J. Davis and Reuben Hersh

All of us are 'drowning in digits' and depend constantly on mathematics for our high-tech lifestyle. But is so much mathematics really good for us? This major book takes a sharp look at the ethical issues raised by our computerized society.

The Blind Watchmaker Richard Dawkins

'An enchantingly witty and persuasive neo-Darwinist attack on the anti-evolutionists, pleasurably intelligible to the scientifically illiterate' – Hermione Lee in the *Observer* Books of the Year

Microbes and Man John Postgate

From mining to wine-making, microbes play a crucial role in human life. This clear, non-specialist book introduces us to microbes in all their astounding versatility – and to the latest and most exciting developments in microbiology and immunology.

Asimov's New Guide to Science Isaac Asimov

A classic work brought up to date – far and away the best one-volume survey of all the physical and biological sciences.

FOR THE BEST IN PAPERBACKS, LOOK FOR THE 🐧

PENGUIN SCIENCE AND MATHEMATICS

The Panda's Thumb Stephen Jay Gould

More reflections on natural history from the author of *Ever Since Darwin*. 'A quirky and provocative exploration of the nature of evolution ... wonderfully entertaining' – *Sunday Telegraph*

Genetic Engineering for Almost Everybody William Bains

Now that the genetic engineering revolution has most certainly arrived, we all need to understand its ethical and practical implications. This book sets them out in accessible language.

The Double Helix James D. Watson

Watson's vivid and outspoken account of how he and Crick discovered the structure of DNA (and won themselves a Nobel Prize) – one of the greatest scientific achievements of the century.

The Quantum World J. C. Polkinghorne

Quantum mechanics has revolutionized our views about the structure of the physical world – yet after more than fifty years it remains controversial. This 'delightful book' (*The Times Educational Supplement*) succeeds superbly in rendering an important and complex debate both clear and fascinating.

Einstein's Universe Nigel Calder

'A valuable contribution to the demystification of relativity' – *Nature*

Mathematical Circus Martin Gardner

A mind-bending collection of puzzles and paradoxes, games and diversions from the undisputed master of recreational mathematics.